Roland Aircraft of WWI

A Centennial Perspective on Great War Airplanes

Jack Herris

Great War Aviation Centennial Series #9

This book is dedicated to my friend and colleague Greg VanWyngarden, whose generosity with photos, information, and advice has been essential to this series.

This book is about aircraft *designed* and built by L.F.G. Roland. Other manufacturer's aircraft built under license by L.F.G. Roland will be covered in the titles on those manufacturers.

Acknowledgements

My sincere thanks to Greg VanWyngarden, Colin Owers, Bill Toohey, Dave Hooper, and Richard Andrews for photographs, Bob Pearson for his color profiles, Aaron Weaver for photographs and image editing, Marty Digmayer and Colin Owers for scale drawings, and the Museum of Flight in Seattle and the Deutsches Technikmuseum in Berlin for photographs. Any errors are my responsibility.

The cover painting by Steve Anderson illustrates the colorful Roland D.VIa flown by *Oblt*. Otto Kissenberth, who was later awarded the *Pour le Mérite*. Many thanks to Wingnuts Wings for their generous permission to use this spectacular painting. Please see Steve's website at: **www.anderson-art.com**

Color aircraft profiles © Bob Pearson. Purchase his CD of WWI aircraft profiles for $50 US/Canadian, 40 €, or £30, airmail postage included, via Paypal to Bob at: **bpearson@kaien.net**

For our aviation books in print and electronic format, please see our website at: **www.aeronautbooks.com**. I am looking for photographs of the less well-known German aircraft of WWI to complete this series. For questions or to help with photographs you may contact me at the email address on the website.

ISBN: **978-1-935881-20-9**

Design and layout: Jack Herris
Cover design: Aaron Weaver
Digital photo editing: Jack Herris, Aaron Weaver

Aeronaut Books

www.aeronautbooks.com

Table of Contents

Right: The statue of Roland in the city of Bremen, erected in 1404. The Roland trade name was inspired by this statue. (Courtesy of Jürgen Howaldt)

Luft-Fahrzeug Gesellschaft (L.F.G.) 'Roland'

The Motorluftschiff-Studien-Gesellschaft mbH (MSG), an aeronautical holding company, established the Luft-Fahrzeug Gesellschaft mbH (LFG) in 1908 to build and market Parseval airships. Headquarters were in Berlin with the production facilities at Betterfeld. A year later MSG purchased the rights to the Wright brothers patents and formed the Flugmaschine Wright Gesellschaft to manufacture Wright Flyers. After building 40 Wright Flyers and training 30 pilots how to fly, the German patent court declared the Wright patents invalid, and Flugmaschine Wright Gesellschaft went into bankruptcy.

LFG entered the aircraft market in 1912 by acquiring the Flugmaschine Wright Gesellschaft. The factory was located at the Johannisthal airfield in Berlin. In early 1913 LFG adopted the corporate trade name Roland, a legendary medieval knight, to avoid confusion with the Luft-Verkehrs-Gessellschaft (LVG) aircraft manufacturer. For their logo LFG used an illustration inspired by the famous statue of Roland erected in Bremen in 1404 to celebrate Bremen's freedom.

Pre-war Roland built a small number of aircraft, with one intention being to supply Germany's African colonies. A Roland *Stahldoppeldecker* (steel-framed biplane) fitted with extra fuel tank captured the world endurance record by staying in flight for 14 hours on 3 February 1914. However, the German air service purchased only two Roland aircraft before WWI, a *Stahl-Pfeil* biplane and a *Stahl-Pfeil Taube*, A.157/13. In December 1914, with the war in its fifth month, Roland received its first military contract for two battleplanes (G.I and G.II) followed by license production of the Albatros B.II(Rol) and C.I(Rol) biplanes. In 1917 the Navy ordered ten license-built Sablatnig floatplanes.

Roland engineers developed the revolutionary

Above: West side of the LFG Roland Kaiserdamm factory.

Below: LFG flight operations at Adlershof, summer 1917. The LFG Adlershof factory burned down on 6 Sept. 1916.

Above: North side of the LFG Roland Kaiserdamm factory.

Below: Great hall of the Kaiserdamm factory before the fire showing the huge space available for the LFG factory.

Wickelrumpf (wrapped body) technology for fuselage construction, which combined layers of wood veneer strips wrapped diagonally around a mold at an angle to each other, and glued and reinforced with fabric. This was patented and used for many Roland aircraft as well as being licensed to Pfalz, who used it for all their subsequent designs. The *Wickelrumpf* technique produced strong, streamlined fuselages

Above: The LFG works at Adlershof after the fire of 6 Sept. 1916.

Below: The fitter's shop in the Kaiserdamm factory.

Above: Woodworking in the Kaiserdamm factory.

Above: Design and drafting area in the Kaiserdamm factory.

Below: Metalworking machines in the Kaiserdamm factory.

Above: Building wood sub-assemblies at Kaiserdamm.

but took a lot of time and labor, so Roland engineers developed a new technique for fuselage construction similar to clinker-planked boat hulls, with the result it was called *Klinkerrumpf* (clinker body). This was also patented and was used for several later Roland designs to save production labor.

On September 6, 1916 LFG's Adlershof factory burned down. However, losses were not as serious as might have been; only seven completed aircraft and ten fuselages were burned. Fortunately, the assembly jigs and fixtures and some drawings were rescued from the fire. *Oblt*. Siegert, *Idflieg* commander, immediately confiscated the empty automobile association building in the Kaiserdamm and what was left of the Roland factory was quickly moved there. The first production aircraft rolled out of the new location on October 9, 1916.

Roland was a major manufacturer and sub-contractor and built 2,859 Army aircraft and perhaps as many as 35 naval aircraft during the war, most built under license to the designs of other companies. This made Roland the fourth largest German WWI aircraft manufacturer, but only 934 aircraft were series production Roland designs.

After the war L.F.G. moved to Seddin, near Potsdam, to produce airplanes, balloons, and life boats. Aircraft construction lasted until 1933, the designs being single-engine passenger aircraft.

LFG Roland Milestones

Above: The LFG *Stahl-Pfeil* (steel 'arrow') biplane of 1913, Roland's first original design, was powered by a 100 hp Mercedes. On 3 Feb. 1914 this aircraft set a world endurance record of 14 hours in the air.

Above: The LFG *Stahl-Pfeil Taube* (steel arrow 'dove') monoplane of 1913 was Roland's second original design.

Note: Most of the photographs on pages 3, 4, 5, & 8 are from *Das Wirken Der Luft-Fahrzeug-Gesellschaft Abteilung Flugzeugbau im Weltkriege* (*The Affect of the LFG Aircraft Division in the World War*) published postwar by LFG.

Below: This Roland D.IIa, completed the morning of 7 Feb. 1917, was the 500th Roland airplane built during the war.

Wickelrumpf

The Roland-designed *Wickelrumpf* wrapped-fuselage construction was also used by Pfalz, who learned it from building Roland fighters under license. This entailed wrapping thin, narrow strips of plywood around a mold and gluing them together into a strong, light, monocoque shell. During the late stages of the production process the fuselage was wrapped by a layer of fabric. Minimum framing was required. The technique facilitated streamlined shapes and conserved scarce resources by using relatively abundant plywood. It also resisted battle damage because the entire shell carried the load, leaving no weak spots in the structure that could be disabled by a single bullet. Although labor intensive, this construction proved so successful that it was used in all known subsequent Pfalz designs, including the Pfalz D.VIII, whose fuselage shells are shown ready for assembly at left.

Captured Pfalz D.III 4184/17 stripped bare for inspection. The light fuselage framework supported an outer plywood shell, created by the *Wickelrumpf* technique.

Above: Building a Pfalz fuselage; minimum framing was needed for the *Wickelrumpf* fuselage shell.

Below and Right: The structure of the second version of the Roland D.III prototype photographed April 20, 1917, after Roland had a production contract, showing its massive, robust center section. *Wickelrumpf* fuselage construction gave great strength without requiring a heavy internal structure.

Frontbestand Inventory of All Roland Designs at the Front

		1914			1915						1916						1917						1918			
Manufacturer and Type		31 Aug	31 Oct	31 Dec	28 Feb	30 Apr	30 Jun	31 Aug	31 Oct	31 Dec	28 Feb	30 Apr	30 Jun	31 Aug	31 Oct	31 Dec	28 Feb	30 Apr	30 Jun	31 Aug	31 Oct	31 Dec	28 Feb	30 Apr	30 Jun	31 Aug
Roland	C.I												35				21	9	1							
	C.II											17	35	37	45	40	53	42	1	1						
	C.IIa													13	11	24			12	1						
	C.III																1									
Roland	D.I														1	7	12	2	1							
	D.II																22	97	41	10	2					
	D.IIa																		128	102	6	3				
	D.III																						9	14		
	D.VIa																								38	58
	D.VIb																								17	12

Above: In the *Frontbestand* table, the Roland C.I is actually the Albatros C.I(Rol) as noted in the main text. The Roland C.III is an error; only one Roland C.III was built and it was destroyed when the factory burned on Sept. 6, 1916. The Roland C.III listed was probably either a Roland C.II or C.IIa.

LFG Roland Factory Scenes

Above: Completed Roland C.II aircraft in front of the LFG factory ready for delivery.

Left: Roland D.I and late (large fin) C.IIa airframes being built on November 8, 1916 in the Kaiserdamm factory.

Above: Transporting a Roland C.IIa to the front by truck from Charlottenburg on November 28, 1916.

Above: Roland D.VI at the firing range at Adlershof testing its machine gun synchronization and aiming.

Roland Two-Seaters

Roland's first aircraft designs, the *Stahl-Pfeil Doppeldecker* and *Stahl-Pfeil Eindecker* built in 1913 were unsophisticated, unarmed two-seat aircraft designed for reconnaissance and observation in Germany's African colonies. Their airframes were primarily build of steel tubing to give them the robustness needed for this demanding role. Aside from their steel airframes, their engines and configurations were typical of German pre-war designs. Few were built and only two were purchased by the German Air Service before the war.

Excepting the G.I, a two-seater to the G-category, the next two-seaters built by Roland were the Albatros B.II(Rol) and C.I(Rol) biplanes built under license. This license production of other manufacturer's designs established a pattern Roland followed until the Armistice; Roland built many aircraft of its own design but produced twice as many aircraft under license.

The streamlined Roland C.II was a landmark design and by far Roland's most successful creation. It was the first Roland design produced in quantity for the German Air Service and for a time was the best German warplane at the front. In addition, the first Roland fighter designs were derived from the successful C.II. Unfortunately, Roland was unable to follow the C.II with another successful two-seater design. As a result, Roland built the Albatros C.X and later the Halberstadt CL.IV under license.

Roland Stahl-Pfeil-Doppeldecker

Above: The Roland *Stahl-Pfeil-Doppeldecker* (Steel-Arrow-Biplane) was a pre-war design; only one was purchased by the German air service. This early aircraft has a 4-cylinder engine; most used more powerful 6-cylinder engines.

Above: Roland *Stahl-Pfeil-Doppeldeckers* had sharply swept-back wings, hence the *Pfeil* (Arrow) designation.

Below: On 3 Feb. 1914 Bruno Langer set a world flight endurance record of 14 hours in a Roland *Stahl-Pfeil-Doppeldecker.*

Above: Roland soon upgraded the engines of *Stahl-Pfeil-Doppeldeckers* to more powerful 6-cylinder types.

Below: Six-cylinder *Stahl-Pfeil-Doppeldecker* sent to Africa before the war for colonial use.

This Page: Additional views of Roland *Stahl-Pfeil-Doppeldeckers.* On 3 Feb. 1914 an aircraft of this type set a world endurance record of 14 hours in the air.

Roland Stahl-Pfeil-Eindecker (Taube)

This Page: Roland *Stahl-Pfeil-Eindeckers* (Steel-Arrow-Monoplanes). Like most German aircraft manufacturers pre-war, Roland built some aircraft to the *Taube* (Dove) configuration. The *Taube* configuration had the significant advantage of being inherently stable in flight, a major consideration at the infancy of aviation. The *Taube* was therefore safer than most other designs of the time, but *Taube* performance was limited by the configuration's high drag. As wartime operations became more demanding, the *Taube* as abandoned for the greater speed, climb, and ceiling of newer biplane designs. Furthermore, the *Taube's* inherent stability and lack of maneuverability was a disadvantage if it was attacked by a hostile aircraft.

Top & Above: Additional views of Roland *Stahl-Pfeil-Eindeckers*. Only one Roland *Taube* was purchased by the German *Fliegertruppe*; it was given the designation A.157/13. Unlike many other *Taube* designs, the Roland had hinged elevators instead of warping elevators.

Left: This uncovered aircraft shows Roland's steel-tube construction.

Roland B.I

The Roland B.I was in fact the Albatros B.II built under license by Roland. Apparently these aircraft were initially designated Roland B.I (870–881/14), indicating that a dozen were built. To reduce confusion *Idflieg* started rationalizing aircraft designations in August 1916, and the Roland B.I was sensibly re-designated the Albatros B.II(Rol) by *Kogenluft* directive on 18 February 1917.

Above: Roland B.I, later designated Alb. B.II(Rol), of *Feldflieger Abteilung* 21 being inspected by *General* von Emmich in the Spring of 1915. Unarmed from the factory, a machine gun has been added at the unit for use by the observer in the front seat.

Right: Roland B.I, later designated Alb. B.II(Rol). An abundance of national insignia were applied to prevent troops shooting at their own aircraft.

Roland C.I

The Roland C.I was in fact the Albatros C.I built under license by Roland. To reduce confusion *Idflieg* started rationalizing aircraft designations in August 1916, and the Roland C.I was rationally re-designated the Albatros C.I(Rol). A total of 88 aircraft were ordered and built as shown in the adjacent table.

Alb. C.I(Rol) Production		
Ordered	**Qty**	**Serial Numbers**
27 Aug. 1915	34	C.640–663/15
31 Oct. 1915	36	C.1800–1835/15
27 Nov. 1915	18	C.1986–2003/15

Left: Roland C.I, later designated Alb. C.I(Rol). The subsequent Roland C.II was the first original Roland design to see mass production.

Below: A standard, Albatros-built Albatros C.I showing its distinctive tail surfaces. The Albatros C.I and Roland C.I [Alb. C.I(Rol)] were powered by either the 150 hp Benz Bz.III or 160 hp Mercedes D.III.

Roland C.II

Above: The LFG Roland C.II was a break-through design for Roland. The exceptional streamlining for the time – it entered service in March 1916 – was developed in conjunction with wind-tunnel testing and full-size testing on a high-speed train. To reduce drag the upper wing was built directly into the fuselage without benefit of cabane struts. The C.II was the smallest, fastest German two-seater when it reached the front, and was faster than most Allied fighters. Initially armed with a single flexible gun for the observer, the second and subsequent production batches included a fixed, synchronized gun for the pilot as shown above. The speed of the C.II coupled with its effective armament made it the first German two-seater that was effective as a two-seat fighter. The relatively high-performance C.II proved a handful for many service pilots and its maneuverability and flying qualities were mediocre. The narrow gap between the wings resulted in airflow interference between them, reducing lift and increasing drag. The view upward was excellent but the view downward and forward was limited, a problem that contributed to numerous landing accidents. *Lt.* Seitz was the pilot of this C.II.

The Roland C.II was a completely new, original design by Roland. When building the Albatros B.II and C.I under license, Roland engineers became concerned about the high drag created by the multitude of bracing wires in these designs. The engineers started exploring methods to reduce drag to increase speed using the same engines currently employed. A wide field of fire for the gunner was also a design priority. This work led to the Roland *Walfisch*, or Whale, as the C.II was nicknamed in Germany.

Much of the engineering was based on the work of *Professor* Ludwig Prandtl at the Göttingen University wind tunnel. Prandtl had taught Roland chief designer *Dipl.-Ing.* Tantzen and consulted on aerodynamics for Roland. Like many significant advances in aeronautics, the Roland C.II combined new construction methods with superior aerodynamics to achieve its jump in performance. In this case, the Roland engineers used their revolutionary *Wickelrumpf* technology, which combined layers of veneer strips wrapped diagonally around a mold at an angle to each other, and glued and reinforced with fabric. This method resulted in a robust, streamlined shell that carried the structural loads. Using this new technology, the Roland C.II was designed by Tantzen, *Ober-Ing.* Cämmerer, *Ing.* Sonntag, and *Ing.* Richter.

Powered by a 160 hp Mercedes D.III engine, the Roland C.II prototype first flew on 24 or 25 October, 1915. Engine failure soon after takeoff damaged the prototype, which had to be repaired. In speed trials the C.II demonstrated a speed advantage of about 30 km/h (19 mph) over contemporary German two-seaters. The excellent speed of the Roland C.II was good enough for an initial production order for 50 aircraft, awarded 23 December 1915.

The type-tests and static load tests were performed in mid-February 1916 on airframes C.4415/15 (work #550) and C.4423/15 (work #558). During testing it was determined that the wing cellule was not strong enough, and an additional drag wire had to be installed. After that minor delay, the first five production aircraft were ready to leave the factory on 7 March 1916.

The streamlined Roland C.II arrived at the front

in March 1916, where it created a sensation, and served on operations until June 1917. At the time, it was the smallest and fastest German two-seater, and certainly looked fast. Its aerodynamic shape and 160 hp Mercedes engine gave it excellent performance for the time. It was as fast as the opposing Nieuport and Sopwith Pup fighters, making it an effective combat aircraft. Because they were so much faster than the other German two-seaters, the Rolands generally operated in groups so as not to be slowed by their contemporaries.

Speed, while especially important in a combat aircraft, is not everything, and the Roland C.II's flight characteristics needed improvement. Probably the most significant short-coming of the C.II was its limited visibility for the pilot forward and downward, making it difficult for pilots to judge their landing flair and making touch-down and landing rollout somewhat tricky and hazardous. Although landing accidents were distressingly common for WWI aircraft, the Roland C.II suffered disproportionately more than its peers.

To improve its combat effectiveness, a fixed, synchronized gun for the pilot was fitted to the C.II starting with the second production batch. The roll-over structure had to be redesigned to accommodate the addition of the gun. Additional changes to the wing structure were made mid-way through the third production batch to improve flight characteristics. The changes, which included moving the interplane I-struts inboard and reducing the wing span slightly, strengthened the wing. It is thought the modified version was designated the C.IIa, whose static test took place in late April or early May 1916. These aircraft reached the front in the summer.

The final significant improvement to the

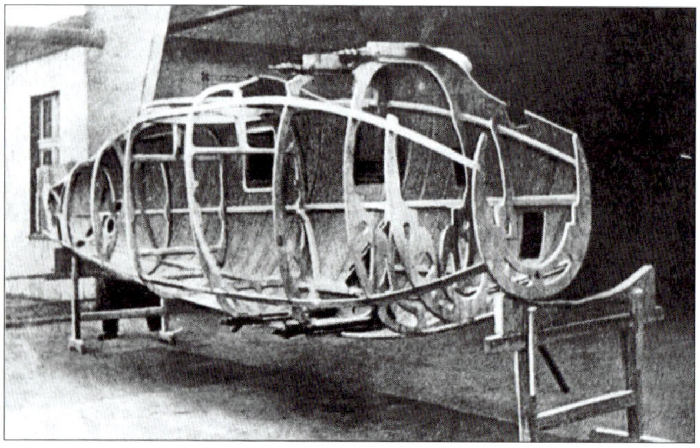

Above: The simple wood frame of the Roland C.II was covered by a streamlined, two-part *Wickelrumpf* shell.

Above: The Roland C.II featured a simple, low-drag landing gear.

Left: The nose of the Roland C.II was very streamlined. Even the exhaust manifold was streamlined. Because the crewmen's heads extended above the wing, the pilot had a turn-over struture for protection.

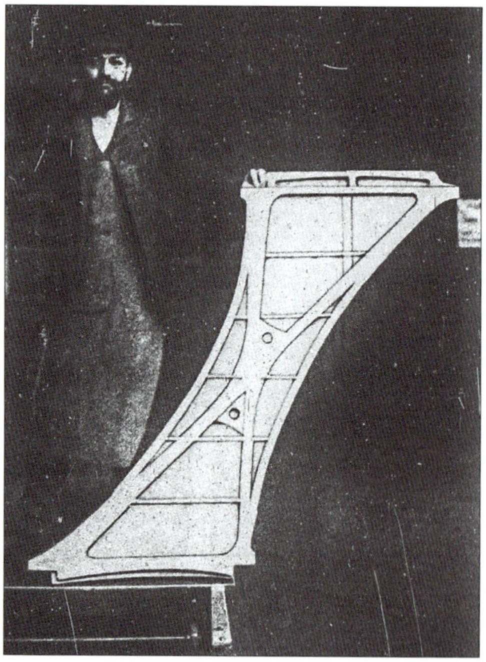

Right: The streamlined interplane strut had internal 'X' bracing connecting the two spars in both wings.

C.IIa was an enlarged vertical tail to improve flight stability and control. The last Roland-built production batch of 40 aircraft, ordered in June 1916, had this feature. The C.II had needed this improvement from the beginning of its service, and the change should have been made during flight-testing before the C.II went into production.

Linke-Hoffman was given an initial production order for the Roland C.IIa(Li) in July 1916 and a final production order for 40 aircraft in September. The type test took place in November 1916 and the production aircraft reached the front in early 1917. By the time these aircraft arrivedat the front new German and Allied aircraft had improved performance; the C.II had lost its performance advantage and its operational service was nearing an end. After being withdrawn from the front, the rugged C.II served with flight schools into 1918.

Roland C.II Specifications (1st Production Version)		
Engine:	160 hp Mercedes D.III	
Wing:	Span	10.33 m
	Chord	1.5 m
	Gap	1.32 m
	Area	27.96 m^2
General:	Length	7.70 m
	Wheel Track	1.75 m
	Empty Weight	789 kg
	Useful Load	520 kg
	Loaded Weight	1309 kg
Maximum Speed:		165 kmh
Climb:	1000m	7 min
	2000m	14 min
	3000m	26 min

Roland C.II Production

Ordered	Qty	Serial Numbers	Designation	Remarks
Dec. 23, 1915	50	C.4413–4462/15	Rol. C.II	One MG, round rollover structure
Mar. 16, 1916	24	C.999–1023/16	Rol. C.II	Two MGs, angular rollover structure
Apr. 16, 1916	45	C.1600–1644/16	Rol. C.II	
Apr. 16, 1916	19	C.1645–1663/16	Rol. C.II/C.IIa	
Apr. 16, 1916	36	C.1664–1699/16	Rol. C.IIa	Aileron control in top wing
June 16, 1916	40	C.2700–2739/16	Rol. C.IIa	Large fin, aileron control in top wing
July 16, 1916	12	C.1851–1862/16	Rol. C.IIa(Li)	Aileron control in top wing
Sep. 16, 1916	40	C.3624–3663/16	Rol. C.IIa(Li)	Aileron control in top wing

Notes: 1. All aircraft after the first production batch had a fixed, synchronized gun for the pilot and angular rollover structures. A total of 267 production aircraft were ordered and built.
2. The production batch ordered June 16, 1916 had the large fin. All others had the original small fin.
3. The first 4 production batches used the ocarina radiator type; radiators varied on the last 4 batches.
4. The first 3 production batches had their aileron controls in the bottom wing. The last 4 batches had the aileron contol in the top wing; location of the aileron contol in the 4th batch is not known.

Right: A Roland C.II of the first production batch as shown by the lack of a fixed gun for the pilot and the rounded rollover cage.

Above: A Roland C.II prototype rests in front of the L.F.G. factory at Adlershof. A similar photo was taken on October 24, 1915, and this photo was probably taken at the same time. An anemomenter airspeed indicator is mounted on the upper left wing for speed measurements. The aileron controls were led through the upper wing but early production aircraft ran the aileron controls through the lower wing. Although all production C.II aircraft used the 160 hp Mercedes D.III engine, at least two prototypes were tested with a 150 hp Benz Bz.III. The *Walfisch* design was patented by *Ing.* R. Richter on October 7, 1915, and the *Wickelrumpf* technology was a closely-held industrial secret until 1918 when publicity photos of the Pfalz factory revealed some details of the process.

Above: The first Roland C.II series production aircraft with company test pilot *Hptm. Freiherr* von Thüna in the cockpit on February 4, 1916. The fin and rudder were slightly increased in area compared to the prototypes, but needed to be even larger for better flying qualities, something finally addressed in the last production batch ordered from LFG. No rollover cage is fitted to this early production aircraft.

Above: The Roland C.II arrived at the front when the comparatively primitive Fokker *Eindeckers* were still the leading German fighters. This early production C.II lacked a fixed, synchronized gun for the pilot so was not yet a two-seat fighter, but it was faster than the *Eindeckers*. There is a great contrast in production technology between the two aircraft.

Below: This early-production Roland C.II shows the clean lines of this break-through design. Eyes and mouth have been painted on to emphasize the *Walfisch* design's resemblence to a fish. Unlike the prototypes, the early production aircraft had their aileron cables routed through the lower wing, then rising upward externally to move the ailerons. The streamlined 'ocarina' exhaust was a distinctive feature of many C.II aircraft.

Above: This early-production C.II is being readied for flight. Wooden planks prevented getting bogged down in mud.

Right: This early-production Roland C.II has been painted with a number of dark (black?) circles for identification.

Right: The streamlined C.II was a notable advance over contemporary two-seaters. The care taken to minimize drag gave it a speed 30 km/h (19 mph) faster than other German two-seaters when it reached the front.

Roland C.II *Meerkatze* of *FFA* 18.

Roland C.II of *Kasta* 8, *Kagohl* 2.

Roland C.II of *Kasta* 8, *Kagohl* 2.

Roland C.II of *Kasta* 8, *Kagohl* 2. This aircraft may have been flown by Manfred von Richthofen because he was in that unit when this aircraft served there.

Above & Left: This streamlined early-production C.II made a good subject for a photo that was made into the Sanke card at left.

Left: This early-production C.II has a dark (black?) band around the fuselage and on the fin for identification. A rack under the observer's window carries signal flares. The Roland C.II was a very photogenic airplane compared to its peers and represented a significant performance increase.

Above: *Kasta* 8 of *Kagohl* 2 lineup with early-production C.II aircraft on the right and other aircraft on the left.

Below: Another photo of the early-production C.II aircraft of *Kasta* 8 of *Kagohl* 2. The C.II in the center of the lineup appears to be the aircraft with black circles shown on page 22.

Right: Portrait of an early-production C.II aircraft with the flight crew, *Oblt*. Maximilian von Cossel at left, and *Vzfw*. Rudolf Windisch second from left, either just before or after their famous raid on a Russian railway line. The ground crew is on the right.

Roland C.II of *Schusta* 27.

Roland C.II flown by *Hptm*. Eduard *Ritter* von Schleich of *Schusta* 28 in original configuration.

Roland C.II flown by *Lt*. Otto Czermack of *Schusta* 28 after a forward-firing gun was fitted in a crude mount.

Roland C.II with shark mouth of *Flieger-Abteilung* 62.

This early-production Roland C.II of *FFA* 6 was flown by *Lt.* Konrad von Bülow and *Oblt.* Tamm. The windows in the fuselage allowed more light into the cockpits and could also be used as emergency escape hatches in case the aircraft turned upside down from a bad landing or being shot down.

Above: Early-production C.II *Meerkatze* of *Feldflieger-Abteilung* 18; the observer has a Lewis gun.

Below: This early-production C.II of *Lt.* Rudolf Windisch of *Feldflieger-Abteilung* 62 has unit-applied colors painted on the fuselage, upper surfaces, and interplane struts for camouflage.

Above: Another photo of early-production C.II *Meerkatze*, which means a green or long-tail monkey.

Left: An early-production C.II aircraft prepares for flight on a snow-covered airfield with the obligatory dog in the foreground.

Above: This unarmed early-production C.II aircraft is in training service. The windows have been modified to slide down to open.

Below: Perhaps in training service, this C.II has its propeller and engine covered by canvas for protection.

Above: Roland C.II of *Feldflieger-Abteilung* (A) 227 surrounded by the ground crew. *Lt.* Stuhldreher sits on the top wing and *Lt.* Wilhelm Allmenröder sits on a wheel.

Below: Later production C.II with fixed gun for the pilot and bomb racks under the center section.

Right: Four mid-production C.II aircraft dismantled for transportation. The fixed machine gun for the pilot was fitted to all but the first production batch and required the rollover structure to be redesigned. The ear radiators, clearly visible here, detracted from the streamlined fuselage lines.

Above: A mid-production C.II in the snow displays its dark camouflage. Unfortunately, the unit and crewmen are unknown.

Right: This closeup of a camouflaged, mid-production C.II shows the crew and armament. The observer has signal flares mounted in a rack aft of his window and the pilot has a rear-view mirror mounted on the turnover pylon.

Above: This closeup of the mid-production C.II shown in full on page 17 shows the armament details. The observer's window has been modified to open by sliding it aft. *Lt.* Seitz was the pilot.

Above: A portrait of *Flieger-Abteilung (Artillerie) 292 (*Artillery cooperation aviation unit 292) with two Roland C.II aircraft in the lineup behind the aircrew. The aircraft on the left is an AEG C.IV, and in the center is an LVG C.IV. Like the Roland C.II, the AEG C.IV used the 160 hp Mercedes D.III. The LVG used the rare 220 hp Mercedes D.IV engine, an 8-cylinder engine created by adding two more cylinders to the 6-cylinder Mercedes D.III. Most German two-seater units were equipped with a variety of types optimized for the different roles they performed.

Above: The aircrew of this mid-production C.II have retained their sense of humor at the front and decorated their aircraft to suit themselves.

Above: This view of a Roland C.IIa(Li) shows the two-color Linke-Hoffman factory camouflage scheme applied to these aircraft. The light-colored rudder and fin are either white or light blue, a signature of C.II aircraft built by Linke-Hoffman.

Above & Below: The mid-production Roland C.II above carries the light, single-color scheme originally used in production. Later production aircraft had the two-color factory camouflage scheme like the C.IIa shown below.

This Page: The mid-production Roland C.II 's shown carry the light, single-color scheme originally used in production.

Below: Wireless installation in the front of the observer's cockpit of a Roland C.II.

Above: Roland C.II aircraft of *Kampfstaffel* 36 wear the unit's chevron marking over their factory camouflage.

Below: A Roland C.II is dismantled for transportation. Moving aircraft by train was very common during WWI because it was more reliable than flying the airplanes to their destination.

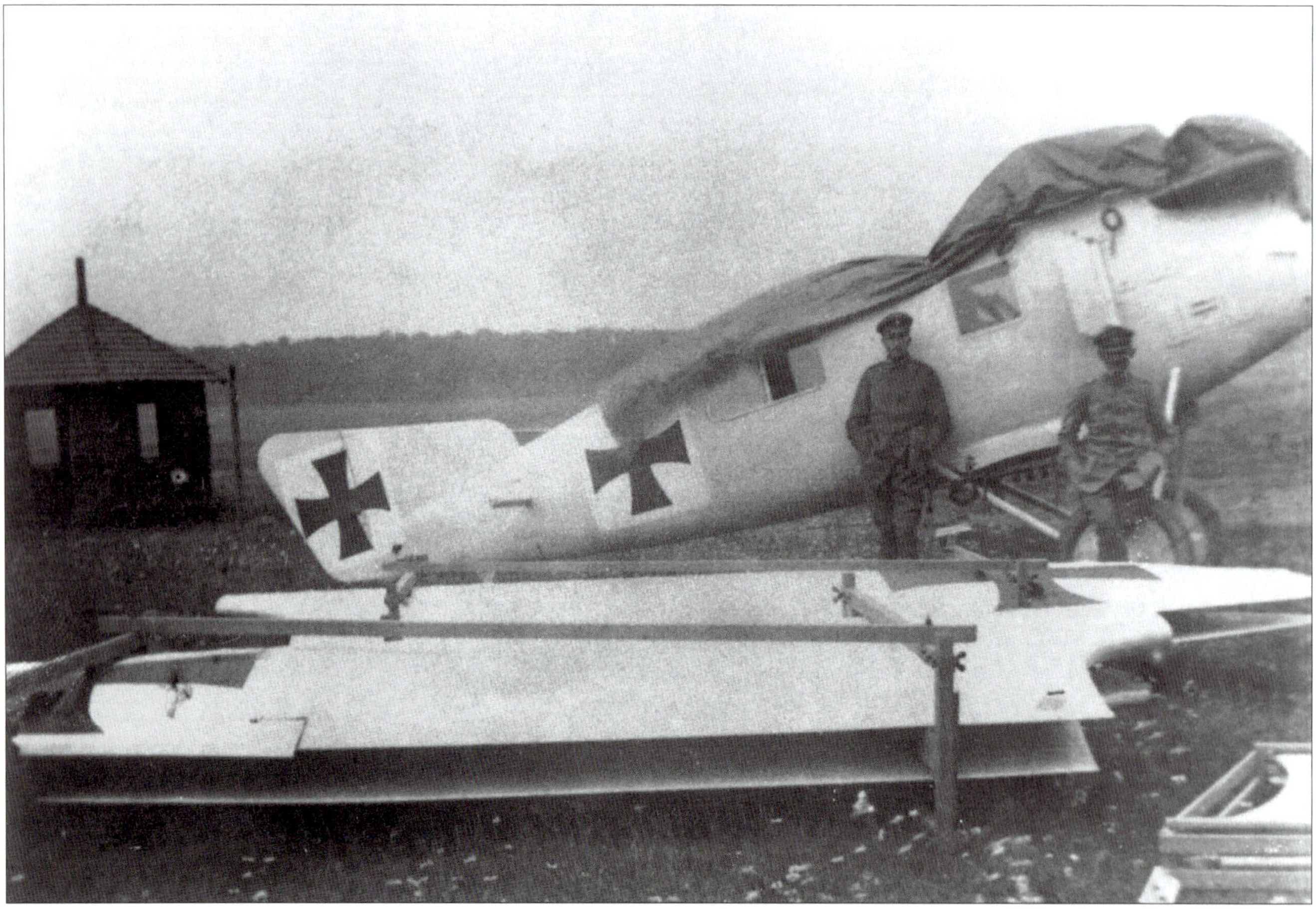

Above: An example of why transportation by train was often preferred; a Roland C.II of Bavarian *Kampfstaffel* 36 has come to grief in during landing. A shark mouth and eye have been painted on the nose and the wheel covers have been quartered, perhaps in Bavarian blue and white. The light interplane strut and lack of a distinct camouflage pattern indicate the dark camouflage may have been applied at the unit.

Below: Flairing for landing too high above ground was not the only way to have a landing accident; this Roland C.II ended its career landing in a canal.

Left: Although many WWI aircraft were destroyed in bad landings, the Roland C.II was notorious for being difficult to land due to the poor field of view forward and downward. This early-production C.II of *Feldflieger-Abteilung* 292b wrecked in a landing accident on April 22, 1916 shows its underside details. The two holes are for the camera and bombs. Although mechanical parts like the engine and machine guns could often be salvaged from a wreck, if the engine was running when the aircraft was wrecked, as evident here by the shattered propeller blades, the crankshaft was typically broken during impact.

Below: Yet another Roland C.II pilot has misjudged his landing. This is graphic evidence of the absolute need for the rollover pylon to protect the crew in case of an accident, although one wonders if the observer survived.

Above: Another view of the same Roland C.II landing accident shown at the top of the facing page.

Above: More Roland C.II landing fun, this time an aircraft of *Schusta* 6.

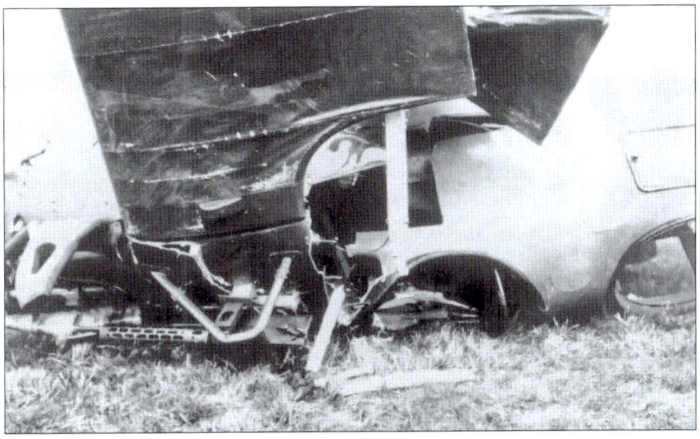

Above: Closeup of yet another Roland C.II landing debacle showing the need for the rollover pylon.

Below: Still more Roland C.II landing fun.

Above & Below: Taxi accidents are more likely when aircraft have no brakes, and most WWI aircraft lacked brakes.

Below: Roland C.IIa(Li) fitted with bomb racks.

Above & Below: Roland C.IIa aircraft of a *Kampfstaffel* lare ined up. The nearest aircraft below has a shark mouth painted on it. The tactical numbers are painted on the vertical fins, and some aircraft have bomb racks. Aircraft "3" through "6" in the lineup have the large tail from the last C.II production batch built by Roland.

Right: An aircrew is photographed with their early-production Roland C.II.

Right: Roland C.II C.1624/16 of the third production batch in embarrassing circumstances. Although *Idflieg* required serial numbers be painted on the aircraft, Roland-built C.II aircraft are notable for not displaying their serial numbers.

Left: A Roland C.II wrecked in a landing accident was hardly a rare event; however, this photo is included to show the interesting fish scales painted on the fuselage. The windows could be used as emergency exits when needed.

Roland C.IIa(Li), unit unknown.

Left: A camouflaged Roland C.II in flight.

Above: This Roland C.IIa(Li) is ready for its next mission. The light-colored fin and rudder on a camouflaged aircraft indicates it was built by Linke-Hoffman, who built only the C.IIa version.

Left: The striking Roland C.II was a favorite backdrop for photos.

Above: A camouflaged Roland C.II of *Flieger-Abteilung* 62 with tactical number '7' painted on its fin and a shark mouth.

Above: A Roland C.II of *Flieger-Abteilung* 62 with shark mouth is photographed with its flight crew in the center (*Oblt.* Maximilian von Cossel at left, and to his left, *Vzfw.* Rudolf Windisch) flanked by its ground crew on the Eastern Front.

This Page & Facing Page: This group of photos depict the Roland C.II aircraft assigned to *Hptm*. Eduard *Ritter* von Schleich and others of *Schusta* 28. The markings of the aircraft are distinctive, as is the fish on the upper left wing in place of an anemometer. An aircraft from the first production batch, it has the early turn-over pylon and no gun for the pilot.

Left: The Roland C.II aircraft assigned to *Hptm*. Eduard *Ritter* von Schleich of *Schusta* 28 attracts a crowd. New aircraft types always draw the curious, and the radically-streamlined Roland C.II was especially popular, gaining a reputation much larger than its small numbers at the front would indicate.

Right: *Hptm*. Eduard *Ritter* von Schleich's Roland C.II in the snow.

Above: *Hptm*. Eduard *Ritter* von Schleich in his Roland C.II.

Above: *Hptm*. Eduard *Ritter* von Schleich's Roland C.II was from the first production batch and lacked a synchronized gun for the pilot.

Above: *Lt*. Otto Czermack's Roland C.II was from the first production batch and lacked a synchronized gun for the pilot. Subsequently a fixed machine gun was mounted to fire above the propeller arc on a crude mount.

Above: *Hptm*. Eduard *Ritter* von Schleich in his Roland C.II.

Left: *Hptm*. Eduard *Ritter* von Schleich's Roland C.II.

Above: *Lt.* Otto Czermack's modified Roland C.II in the snow. A fixed machine gun for the pilot has been added on a crude mount, apparently to fire over the propeller arc, to compensate for the missing pilot's gun installed on later production aircraft.

Below: A dismantled mid-production Roland C.II gets a ride to its next airfield.

Roland C.II of *Lt*. Rudolf Windisch of *Feldflieger-Abteilung* 62; it has unit-applied colors painted on the fuselage, upper surfaces, and interplane struts for camouflage.

Roland C.II with shark mouth of *Flieger-Abteilung* 62.

Roland C.IIa with large tail, *Schusta* 27.

Roland C.IIa with large tail of an unknown unit.

Above: The final Roland production batch of large-tail C.II aircraft under consturction at the Kaiserdamm factory.

Below: Mid-production Roland C.II in flight.

Above: An early production Roland C.II with armament supplemented by a captured Lewis on an improvised mount.

Above & Above Left: Roland C.II aircraft of the first Roland production batch.

Left: An early production Roland C.II shows its clean lines.

Above: Roland C.IIa of *Schutzstaffel* 27 crashed May 27, 1917. This aircraft is from the final Roland production batch and finally has the large fin needed for greater stability, something the aircraft needed from the beginning. It is camouflaged and has a chevron in front of the fuselage cross; the numeral '1' is in front of the chevron.

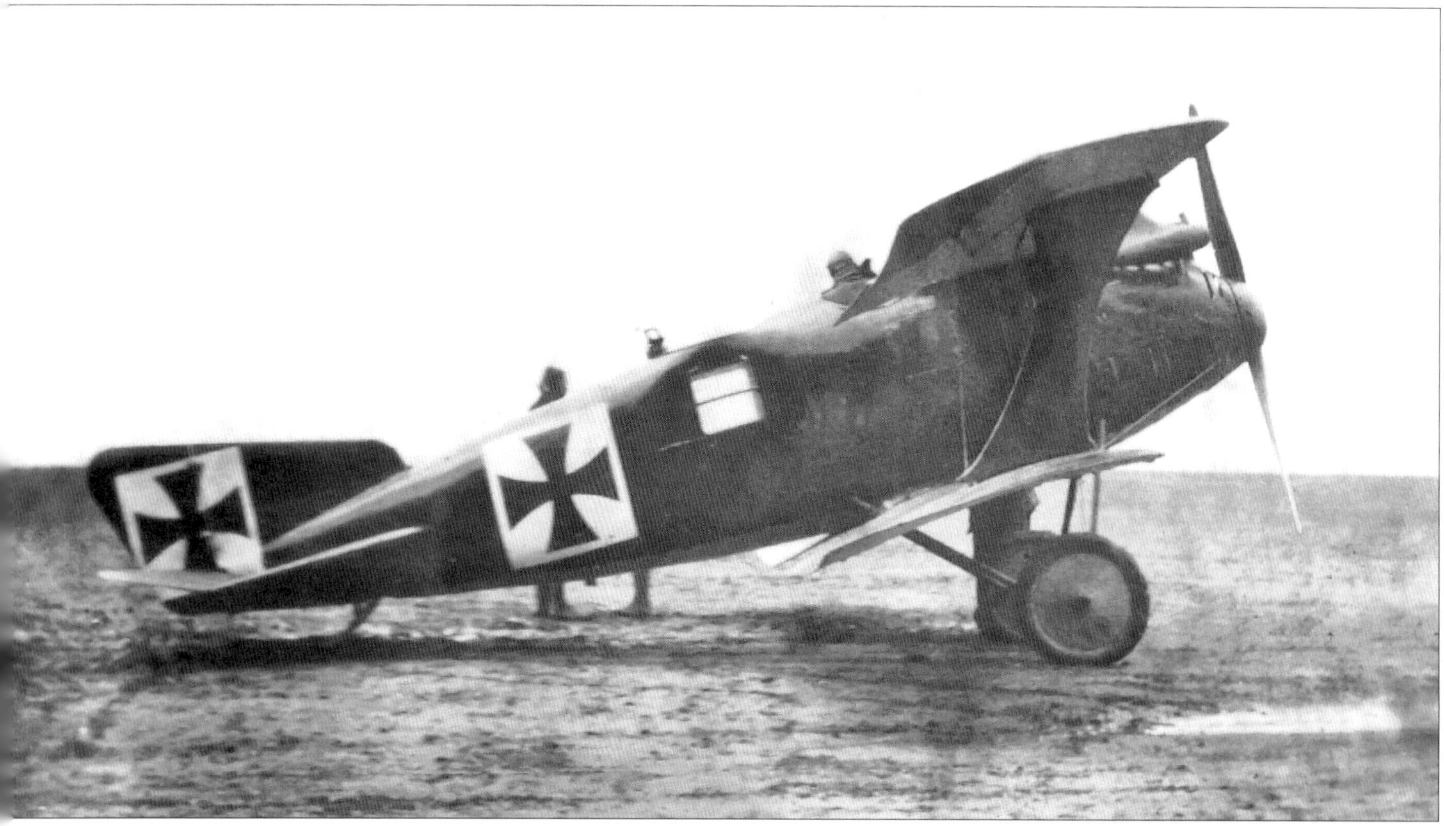

Above: Mid-production Roland C.II. The Roland C.II was by far the most famous and significant Roland design. Not only is it the only well-known Roland design, for a time it was the best German aircraft at the front. Its innovative streamlining immediately made other two-seaters obsolete and its patented fuselage structure was used by Pfalz (under license) for all new Pfalz aircraft designed from the time Pfalz built Roland fighters under license.

Above: Roland C.IIa from the final Roland production batch with the large fin needed for greater stability. It is camouflaged and the pilot's machine gun has been raised well above the normal position.

Above: Large-tail Roland C.IIa 2701/16 of *Feldflieger-Abteilung (A)* 292b in the spring of 1917 at Houplin airfield in northern France. It has had the aft fuselage repaired; external stringers for additional strength have been added to the aft fuselage. The pilot is *Lt*. Otto Fuchs, who later became a fighter pilot and served with *Jastas* 30, 35b, and 77b. The observer is his brother *Lt.d.R.* Rudolf Fuchs, who later became a fighter pilot who served with *Jastas* 77b and 35b.

Right: Roland C.II with *Lt*. Boes in the observer's cockpit and *Oblt*. Dostler at right. *Lt*. Boes is complaining to the *Waffenmeister* Lademacher at left about gun stoppage problems.

This Page: The Roland C.II at upper right flown by Otto Czernak was equipped with an interesting instrument in the cockpits that enabled the observer to send directions to the pilot. Above left, the pilot's receiver is shown behind the control wheel, and below is a closeup of the sender in the observer's cockpit. The German text and apparent meanings are: ***Pumpen*** – Pump (the elevators up and down?), text obscured by arrow, ***Feind*** – Enemy, ***Gleitflug*** – Glide or cruise slowly, ***Gut*** – Good (apparently sent when the pilot was following directions), ***Links*** – To the left, ***Hafen*** – Head for home, ***Rechts*** – To the right, ***Drückn*** – Dive.

Above & Right: A Roland C.II and a visiting elephant from Hagenbeck's *Tierpark* in Hamburg during a visit to the *Flieger-Beobachterschule* Thorn on 17 August 1917. Hagenbeck's *Tierpark* (Zoo) was world famous and would tour units for morale purposes with some of their animals.

Above: A large-tail Roland C.IIa of the last Roland production batch with tactical number '3'.

Right: A large-tail Roland C.IIa of the last Roland production batch.

Below: A Roland C.II with standard tail and factory camouflage and markings for mid-production.

Roland C.III

The Roland C.III was a development of the C.II with conventional two-bay interplane struts replacing the streamlined, single-bay I-struts of the C.II. While the I-struts aided streamlining, they also obscured the view to the sides and the conventional struts improved the view for the crew. The C.III also was fitted with the more powerful 200 hp Benz Bz.IV engine typically used in second-generation C-types like the DFW C.V, etc. The C.III remained a single prototype and was destroyed when LFG's Adlershof factory burned down on 6 September 1916.

Roland C.III Specifications	
Engine:	200 hp Benz Bz.IV

Above & Below: Clearly derived from the C.II, the Roland C.III had larger wings with conventional two-bay interplane bracing. Intended for the more powerful 200 hp Benz Bz.IV engine, it appears like a 160 hp Mercedes D.III is installed here, perhaps for initial flight testing. Armament was the standard fixed Spandau for the pilot and flexible Parabellum for the observer. The sole prototype was destroyed when LFG's Adlershof factory burned.

Roland C.IV

There is no known information for the Roland C.IV. Because the Roland-built Albatros C.X was produced about the same time *Iflieg* rationalized the aircraft designation system, the Albatros C.X(Rol) could originally have been designated Roland C.IV, C.VI, or C.VII in company records. In any case, apparently there is no original Roland design built that was designated Roland C.IV. Later Roland produced other two-seaters under license, the Hannover CL.II(Rol) and Halberstadt CL.IV(Rol). These were too late to receive a confusing designation under the old system, but perhaps Roland management assigned a Roland designation to them, or, more likely, these remained unbuilt Roland projects.

Roland C.V

The Roland C.V was a compact, streamlined two-seater derived from the Roland D.II fighter. Like the D.II the C.V was a single-bay biplane that utilized *Wickelfumpf* technology for the fuselage structure. The C.V was powered by a 160 hp Mercedes D.III engine and had one fixed and one flexible machine gun. Apparently there was no significant advantage over the C.II and only one C.V was built.

Roland C.V Specifications	
Engine:	160 hp Mercedes D.III

Above: The Roland C.V was derived from the D.II fighter. It had the same 160 hp Mercedes D.III engine used in the D.II and larger wings with conventional single-bay interplane struts. Armament was the standard fixed Spandau for the pilot and flexible Parabellum for the observer. Only one Roland C.V was built.

Roland C.VI & C.VII

See the Roland C.IV for a brief discussion of the unknown Roland C-types. These were likely un-built Roland designs but may have been factory designations for C-types built under license.

Roland C.VIII

The Roland C.VIII was a large, two-bay two-seater with conventional struts. Clearly derived from earlier Roland designs that utilized *Wickelfumpf* technology for the fuselage structure, it was a competitor for the Albatros C.X and was similarly powered by the 260 hp Mercedes D.IVa or Maybach Mb.IV. After evaluation of these types Roland was ordered to build the Albatros C.X, itself a mediocre aircraft, under license instead of its own C.VIII design. The C.VIII was the last original Roland two-seat design to be built; after the C.VIII Roland built other two-seat types under license.

Roland C.VIII Specifications	
Engine:	260 hp Mercedes D.IVa or 245 hp Maybach Mb.IV

Above & Left: The LFG Roland C.VIII was a bulky, ungainly design. It featured coventional two-bay wings with ailerons on all wings; the upper and lower ailerons were connected by an actuating strut. It was the last original C-type design built by Roland, who were directed to build the Albatros C.X under license instead of their own C.VIII.

Roland Fighters

The immediate success of the Roland C.II quickly lead Roland to develop a single-seat fighter based on the technology used for the C.II. The resulting Roland D.I shared the 160 hp Mercedes D.III engine from the C.II along with its general design.

Performance of the Roland D.I was very good for the time, and the aircraft was sturdy. Unfortunately, the pilot's field of view forward and downward were severely obstructed, necessitating a quick redesign to improve visibility. The resulting Roland D.II was an insuffcent improvement and another redesign was needed to finally solve the problem, resulting in the Roland D.III.

The problem with these early Roland fighters was that they inherited the problems of the C.II, mediocre maneuverability and handling qualities, along with its strengths of high speed and robust construction. The competitive Albatros series of fighters had similar speed but better handling and maneuverability, with the result that the Albatros fighters were far more popular and built in far larger quantity. Additional Roland development of the D.IV, their first triplane fighter, and the D.V, a development of the D.III, also lead to a dead end.

Roland perservered and finally designed a new fighter, their D.VI, with good maneuverability and handling qualities. Just as it seemed Roland might be successful with its fighters, other competitors, espeically a newly-resurgent Fokker, designed better aircraft. The good Roland D.VI was over-shadowed by the exceptional Fokker D.VII.

During the final year of the war Roland devoted significant resources to fighter design, but to no avail. Other German aircraft manufacturers were doing the same thing and some were more successful at it. Fokker was especially successful, and Pfalz also surpassed Roland in general-purpose fighters. Moreover, Junkers and Zeppelin were building all-metal fighters that were far advanced over any Roland designs. As a result, Roland never became famous for building fighters; the C.II was the only famous Roland design of the war.

Above: The LFG Roland D.I prototype was the first Roland fighter design and direct ancestor for all early Roland fighters. Clearly derived from the Roland C.II, it used the same engine and technology that made the C.II so successful and shared the strengths and weaknesses of the C.II. The upper wing was built directly into the fuselage without benefit of cabane struts, limiting the pilot's view forward. This feature required two re-designs to provide adequate forward view. Unfortunately, the mediocre maneuverability and handling qualities the D.I inherited from the C.II were a more severe limitation for a single-seat fighter than a two-seater and were not overcome until Roland developed the Roland D.VI.

Roland D.I

In 1916 LFG initiated the design of a new fighter plane under the direction of *Dipl.-Eng.* Tantzen. The new fighter was derived from the earlier Roland C.II two-seater; it used the same 160 hp Mercedes D.III engine, the same *Wickelrumpf* "wrapped fuselage" construction, and similar aerodynamic and structural technology.

Three prototypes were ordered in April 1916. In August 1916 *Idflieg* completed testing of the D.I and awarded LFG a contract for 60 fighters. At the same time Pfalz was awarded a contract to produce 20 Roland D.I fighters under license. These aircraft were initially known as the Pfalz D.I, but later, after *Idflieg* rationalized the aircraft designation system, these aircraft were known as the Roland D.I(Pfal).

The Roland D.I had two synchronized LMG 08/15 machine guns buried in the nose. This gave excellent streamlining but made them nearly inaccessible to the pilot in case of a malfunction, common at the time. The D.I arrived at the front in October 1916, but a fire at the LFG factory in 6 September reduced production to a trickle at this important time.

The Roland C.II was known as the *Walfisch* (Whale) and the smaller D.I was nicknamed *Haifisch* (Shark), but this was not an official name. Like the C.II, the D.I was strong and fast, but lacked the maneuverability needed for a first-rate fighter. In addition, the pilot's field of view forward and downward was compromised by the design, so few Roland D.I fighters were built.

Roland D.I Specifications

Engine:	160 hp Mercedes D.III	
Wing:	Span	8.90 m
	Chord (upper)	1.45 m
	Chord (lower)	1.45 m
General:	Length	6.93 m
Maximum Speed:		180 kmh
Climb:	1000m	4 min
	2000m	8 min
	3000m	13.5 min
	4000m	22 min
	5000m	28 min

Roland D.I Production

Ordered	Qty	Serials	Notes
April 1916	1	112/16	Prototype
Aug. 1916	60	830–889/16	1st production
Sept. 1916	20	1680–1699/16	Pfalz-built
Oct. 1916	40	Unknown	See note
Note: May have been changed to D.II			

Below: Prototype Roland D.I 112/16 at the factory. The photo has been re-touched to outline the shape of the vertical tail surfaces which did not contrast much with the sky. The upper wing was built directly into the fuselage without benefit of cabane struts, limiting the pilot's view forward and downward. Unlike the Roland C.II, I-type interplane struts were not used. It was the first of three prototypes ordered in April 1916, the other two were built as Roland D.II fighters.

Above: The production LFG Roland D.I shown here at Adlershof was also built under license by Pfalz as the Pfalz D.I. The crosses are of normal proportions, which indicates this may be a Pfalz-built aircraft; Roland-built aircraft had thicker crosses. Production D.I fighters did not have the fuselage side window present in the D.I prototype.

Below: This Roland-built D.I in typical factory sprayed camouflage has had an accident. The angular roll-over structure of the D.I resembled that used in later-production C.II aircraft and was not used in later Roland fighters.

Roland D.II & D.IIa

The Roland D.II was a development of the D.I designed to improve the pilot's field of view forward and downward. The upper wing was now mounted on a small pylon and the side-mounted 'ear' radiators of the D.I were replaced with an airfoil radiator mounted in the upper wing. Both changes improved the pilot's field of view forward and downward. Engine and armament remained the same as the D.I, a 160 hp Mercedes D.III engine and two synchronized LMG 08/15 machine guns buried in the nose.

Despite these improvements, the maneuverability and flying characteristics of the D.II were no better than the D.I. The competing Albatros fighters were somewhat easier to fly and more maneuverable than the Rolands and were much more successful in combat.

Regardless, LFG received a contract for 30 D.II fighters in October 1916. In November Pfalz was awarded a contract for 100 of these fighters as the Pfalz D.II, later re-designated Roland D.II(Pfal).

When the somewhat more powerful 180 hp Argus As.III engine became available, it was used in the D.II to create the D.IIa. This had the additional benefit of freeing up more Mercedes engines for the production of Albatros fighters. The Roland D.IIa was ordered in greater quantity than the D.I; 40 were ordered in November 1916, 100 in January 1917, and a final 50 in March 1917.

The D.IIa had an overall performance similar to the D.II with Mercedes engine, but the Argus engine experienced a significant power loss at higher altitudes, significantly reducing the rate of climb of the Roland D.IIa above 3,000 meters. This was a serious limitation of the D.IIa as a fighter and it was never as successful as the Albatros fighters.

Roland D.II/IIa Production

Ordered	Qty	Serials	Notes
April 1916	2	113–114/16	Prototypes
Oct. 1916	30	2800–2829/16	D.II
Nov. 1916	40	3701–3740/16	D.IIa
Nov. 1916	60	2830–2929/16	D.II & D.IIa, Pfalz-built
Nov. 1916	40		
Jan. 1917	100	150–199/17, 450–499/17	D.IIa
Jan. 1917	100	300–399/17	D.IIa, Pfalz-built

Above: Pfalz-built Roland D.II on the Eastern Front. From building Roland fighters Pfalz learned the *Wickelrumpf* wrapped-plywood fuselage construction that was used in all subsequent original Pfalz fighter designs. The Roland D.II replaced the ear radiators of the D.I with airfoil radiators for lower drag, and mounted the upper wing on a pylon for better visibility.

Roland D.II Specifications

Engine:	160 hp Mercedes D.III	
Wing:	Span (upper)	8.90 m
	Span (lower)	8.50 m
	Chord (upper)	1.45 m
	Chord (lower)	1.45 m
	Gap	1.35 m
	Area	22.78 m^2
General:	Length	6.93 m
	Height	3.11 m
	Empty Weight	635 kg
	Loaded Weight	815 kg
Maximum Speed:		180 kmh
Climb:	1000m	3 min
	2000m	6 min
	3000m	10 min
	4000m	15 min
	5000m	23 min

Roland D.IIa Specifications

Engine:	180 hp Argus As.III	
Wing:	Span (upper)	8.90 m
	Span (lower)	8.50 m
	Chord (upper)	1.45 m
	Chord (lower)	1.45 m
	Gap	1.35 m
	Area	22.78 m^2
Maximum Speed:		180 kmh
Climb:	1000m	3 min
	2000m	6.5 min
	3000m	11 min
	4000m	18 min
	5000m	29 min

Below & Bottom: Roland D.II(Pfal) 2872/16 photographed in March 1917 before it was camouflaged. The gravity fuel tank in the left wing was used on late-production aircraft.

Above: The Roland D.IIa had a slightly lengthened nose to accommodate its 180 hp Argus As.III engine. Just below the aileron control rod is the Roland logo.

Below: Roland D.IIa serving at the front.

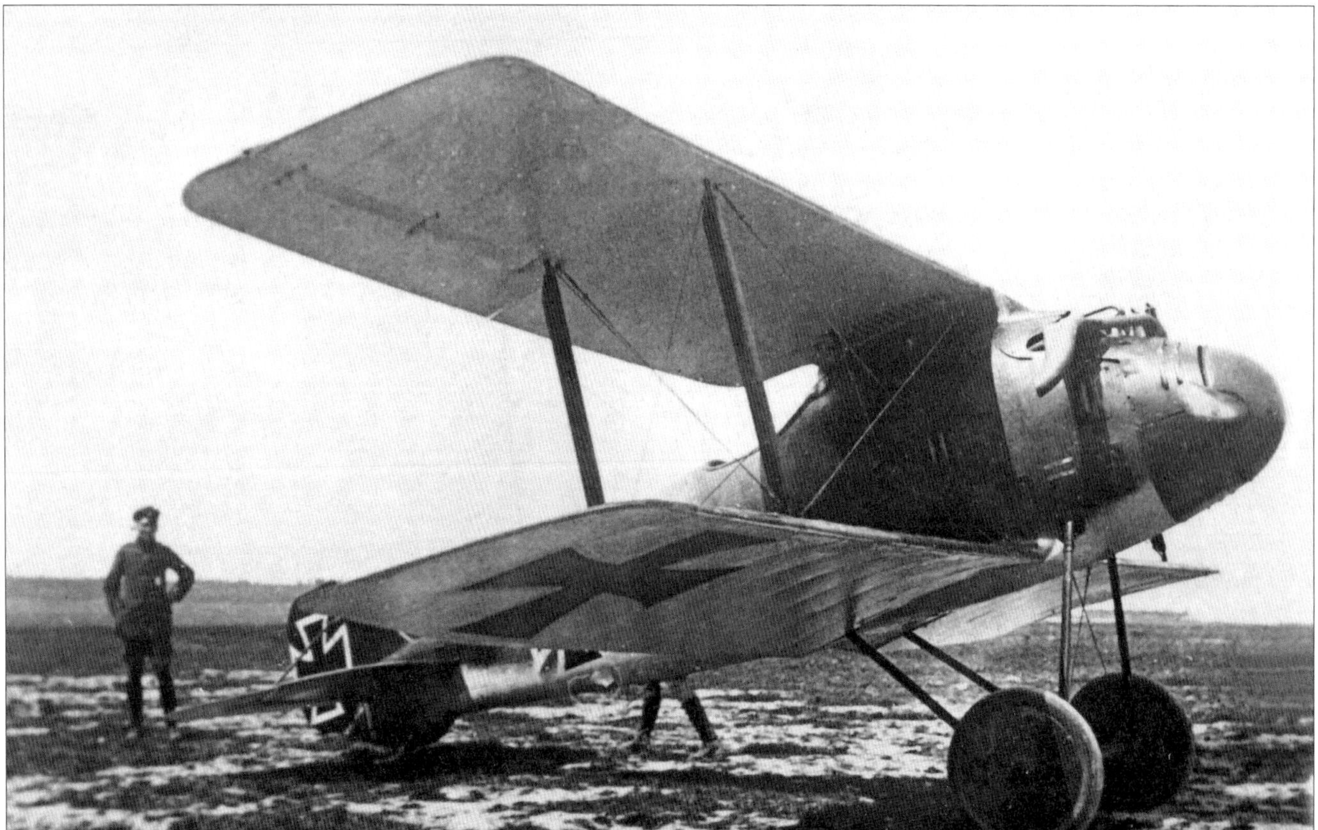

This Page: Three views of the prototype Roland D.II taken on Oct. 22, 1916 in front of the LFG flight test hangar. An anemometer airspeed indicator is mounted on the upper left wing.

Above: This Roland D.IIa interests the visiting officers.

Below: Roland D.IIa fighters in storage waiting distribution to front-line units.

Above & Below: Roland D.IIa(Pfal) 374/17 at Koledziany airfield on the Eastern Front. In addition to the standard insignia, the lack of a tail skid fairing identifies this aircraft as Pfalz-built. The Roland D.IIa was used extensively on the Eastern Front, where the air combat environment was more permissive than on the Western Front and the D.IIa could make use of its strengths, good speed, strong armament, and a robust structure, against two-seaters while its weaknesses, poor maneuverability and handling and modest climb over 3,000 m due to its Argus engine, were not as important.

Above: This unit lineup features a Roland-built D.II in the foreground followed by Halberstadt and Fokker biplanes.

Above: Roland D.II *Edith* and Albatros D.III fighters of *Jasta* 15.

Above: At 44 years old , *Lt*. Friedrich N. Güttler was one of the oldest fighter pilots and was, of course, known as 'Papa' to the other pilots of his unit, *Jasta Ober Ost* on the Eastern Front. Here he is in the cockpit of his Roland-built D.IIa.

Above: 19-victory ace *Vzfw*. Gerhard Fieseler in the cockpit of his Roland D.IIa of *Jasta* 25 in Macedonia. A small windscreen is fitted on each side of the central pylon; the one at right has a small hole for aiming the guns. The bars in front of the windscreens mounted the rear gunsights. The large instrument is the altimeter.

Above: Crashed Roland D.IIa of *Jasta* 25 in Macedonia; the abashed pilot is photographed with the evidence.

Above: Roland D.IIa "K" of *Jasta* 25 in Macedonia is ready for take-off.

Above: There was at least one in every unit; Pinkert of *Jasta* 25 in Macedonia does a head-stand in his Roland D.IIa.

Above: Lineup of Roland D.IIa fighters of *Jasta* 25 in Macedonia. The individual letters were the first initials of the pilots' last names. All are Roland-built except the last in line appears to be built by Pfalz.

Below: Roland D.IIa "S" of *Jasta* 25 in Macedonia is inspected by a group of curious officers.

Above: 19-victory ace *Vzfw.* Gerhard Fieseler in the cockpit of *Lt.* Ahelfeld's Roland D.IIa of *Jasta* 25 in Macedonia. *Jasta* 25 Roland fighters were normally identified by the initial of the pilot's last name; here, 'A' for Ahelfeld. Fieseler's aircraft would most likely have had an 'F' for identification. After the war Fieseler started his own aircraft company; his most famous product was the Fi-156 Fieseler *Storch* (Stork) known for its exceptional low speed flying qualities and ability to make very short landings and take-offs.

Above: *Jasta* 27 lineup; a single Roland D.II is at left in a lineup consisting mostly of Albatros D.III and D.V fighters.

Roland D.I of *Jasta* 5.

Roland D.II prototype.

Roland D.II of *Jasta* 27.

Roland D.IIa of *Jasta* 25 in Macedonia. Most likely this was the aircraft flown by 19-victory ace *Vzfw*. Gerhard Fieseler.

Roland D.II with the coat-of-arms of the city of Halle an der Saale.

Roland D.IIa of *Jasta* 27 flown by *Unteroffizier* Stein.

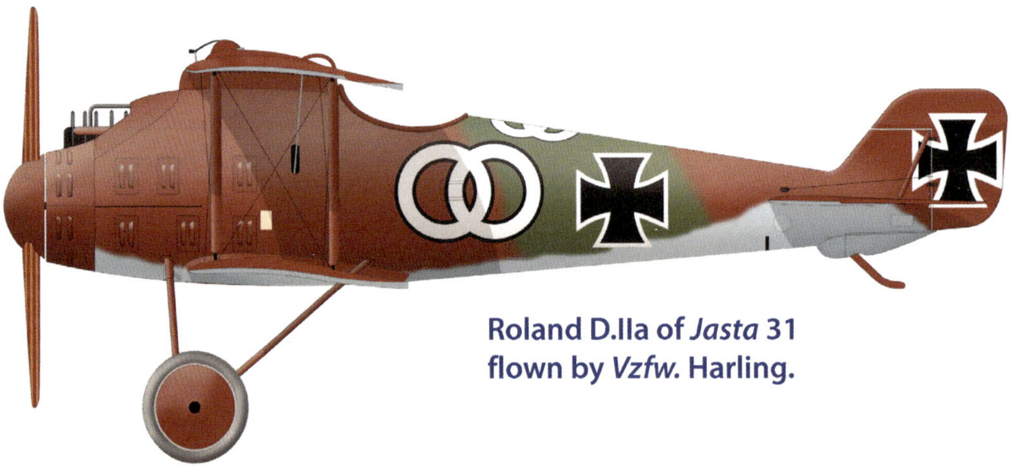

Roland D.IIa of *Jasta* 31 flown by *Vzfw.* Harling.

Roland D.IIa of *Jasta* 32 flown by *Vzfw.* Paul Raetsch.

Above: The port radiator section of this Roland has been removed for maintenance, clearly showing its normal location.

Above: *Jasta* 27 mishap; *Unteroffizier* Stein stalled and crashed from 30 m during takeoff. Despite breaking the tail off his sturdy Roland D.IIa, Stein was not injured in the crash.

Above: Roland D.II with *Lt.* Anslinger.

Above: Roland D.II fighters in wooden hangars with *Flieger-Abteilung* 15 on the Eastern Front. The men appear to be watching a flying display of some sort.

Above: Roland D.II and a Fokker D.II share an airfield.

Above: Roland D.IIa(Pfalz) 713/16 of *Flieger-Abteilung* 21, Eastern Front. Some Pfalz-built aircraft had a faired tail skid.

Above: Roland D.IIa built by Roland wears the standard Roland factory finish devoid of serial numbers. However, data is stenciled below the cockpit. Above the rigging datum line is *Leergewicht* 600 kg (Empty weight 600 kg); below the datum line is *Zul. Nutzlast bei vollem Tank* 167 kg (Permissible useful load with full tank 167 kg).

Below: Roland D.IIa in the foreground and two Mercedes-engined Roland D.II fighters in th background. The two nearest fighters were built by Roland; the fighter in the background has no camouflage on its fuselage and may be Pfalz built.

Above: This Roland D.II fighter carried two broad white fuselage bands as identify markings.

Above: Roland D.II fighters in the field; two of them carry additional identity markings on their fuselages.

Left: The pilots field of view forward and downward from the Roland D.II was little better than that from the C.II with similar results; the landing flair could not be judged well and landing accidents were common.

Roland D.II *Edith* serving with *Jasta* 15.

Roland D.IIa serving on the Eastern Front, unit unknown.

Roland D.II, unit unknown.

Roland D.IIa(Pfal) serving on the Eastern Front, unit unknown.

This Page: Roland D.IIa fighters in the field in Bulgarian service. The aircraft served in typical factory camouflage and markings for German fighters. Fast and well-armed, the Roland D.II/IIa was able to intercept Allied two-seaters with some success but was significantly inferior to the best Allied fighters like the Spad 13 and SE.5a.

Above: Dramatic view of a Roland D.IIa in Bulgarian service in 1918 as indicated by the late-style insignia.

Below: A Bulgarian Roland D.IIa after a landing accident is used as a background for a group photo.

Above: Pfalz-built Roland D.IIa (w/n 539) captured by the Russians after being shot down on Sept. 9, 1917 by *Poruchik* (Lieutenant) Alexander Sveshnikov of the Russian 7th Corps air detachment, 2nd fighting air group. The pilot escaped back to German lines. The broad white marking along the fuselage side is distinctive.

Above: Another view of the aircraft shown above.

Right: Roland D.II being recovered after a pilot's 'head-stand'.

Below: Roland D.II or D.IIa ready for take-off.

Above: Roland D.IIa of *Vzfw*. Paul Raetsch of *Jasta* 32 wears a macabre skull and bones on its spinner.

Below: Roland-built D.IIa after a landing accident shows its underside details. Although embarrassed, pilot *Vzfw*. Paul Raetsch of *Jasta* 32 fortunately appears unhurt. This may be the aircraft also shown above but this is not confirmed.

Roland D.II of *Jasta* 32.

Roland D.II on the Eastern Front.

Roland D.IIa of the *Jastaschule* at Valenciennes.

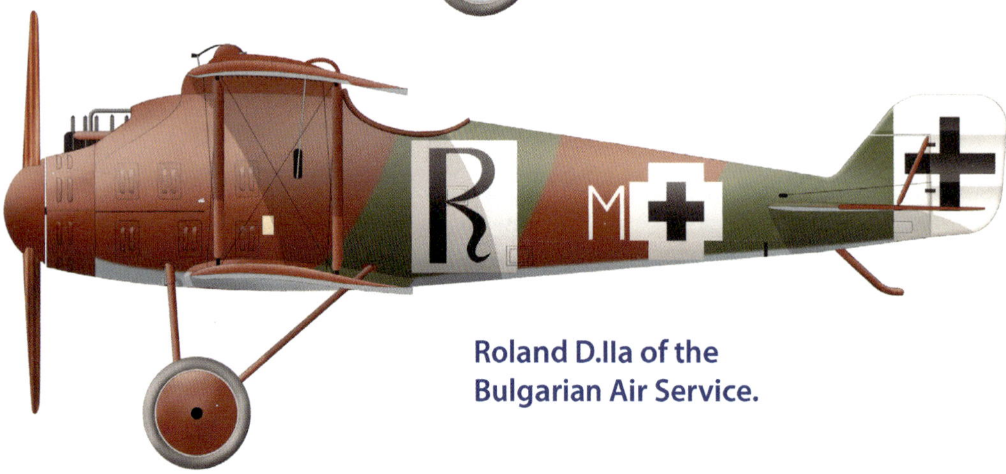

Roland D.IIa of the Bulgarian Air Service.

Above: Roland D.IIa fighters of *Jasta* 25 lined up in Macedonia.
Below: Roland-built D.II or D.IIa prepares for a mission from a snowy airfield.
Bottom: D.II fighters waiting for their next mission.

Above: Typical Eastern Front mixed unit with Roland D.IIs and an Albatros C.III.

Above: Roland D.IIs on the Eastern Front; the aircraft in front was built by Pfalz, and that in the rear was built by Roland. The distinguishing feature is the thickness of the iron cross insignia; the Pfalz-built aircraft has insignia of normal proportions, while Roland painted thicker crosses on the aircraft it produced.

Above: Roland D.II prototype in October 1916 flying over the Adlershof aerodrome.

Above: Production Roland D.II in flight.

Left: Roland D.II or D.IIa number 19 at *Jastaschule* I at Valenciennes.

Roland D.II(Pfal). This aircraft served on the Eastern Front.

Roland D.II(Pfal). This aircraft served on the Eastern Front.

Roland D.II(Pfal). This aircraft served with *Kest* 4 on the Western Front.

Roland D.II(Pfal), unit unknown.

Above: Roland D.II of *Kest* 4b shares an airfield with a Fokker D.II in the background.

Above: French postcard of von Hausen's downed Roland D.II captured May 16, 1917.

Above: French postcard of the downed Roland D.II of *Uffz.* Tolischuss, *Kest* 3, made POW May 23, 1917.

Below: The crest on the side of this Roland D.IIa is the *Municipal Wappen* (coat-of-arms) of the city of Halle an der Saale in southern Germany, in what is now Saxony-Anhalt. The stars and crescent moon were red on a white shield.

Above & Below: Roland D.IIa number 17 at *Jastaschule* I at Valenciennes. The number was painted yellow.

Above: The Mercdes D.III engine removed from a D.II reveals details of the Roland's structure.

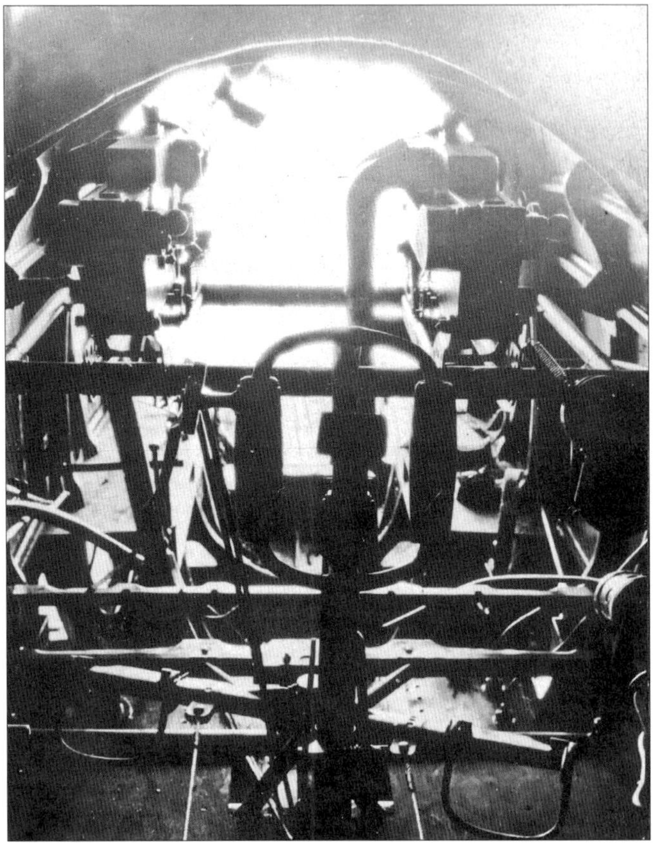

Above: View from the cockpit of a D.II with engine removed shows details of the structure and guns.

Left: This Roland D.II(Pfal) with "J" marking behind the insignia is in an embarrassing situation. One wonders how the pilot could find this building so precisely when it is surrounded by so much open terrain.

Below: These Roland-built D.II fighters have been disassembled for transportation. They are in the gun pits having their guns sighted and tested.

Right: Roland D.II in flight. If the Roland D.II was not the superior fighter the Albatros was, it was a welcome replacement for the primitive *Eindecker* fighters on the Eastern Front. The Roland was faster, better armed, and stronger than the Fokker and Pfalz *Eindeckers* and was significantly superior to them. It was also stronger than the V-strutter Albatros fighters and was not subject to the lower-wing structural failures that plagued those fighters.

Left: This well-used Roland D.II(Pfaz) appears to be at a fighter school. The numeral "8" appears below the cockpit and the lower left wing has been recovered in new fabric after a repair.

Below: This Roland D.II(Pfaz) has had some of its cowling panels removed for engine maintenance. An Albatros C.III is in the background.

Above: This Roland D.II or D.IIa carries a checkerboard personal marking.

Below: This Roland D.II, possibly of *Jasta* 32, carries a personal marking of two white birds on each side of the fuselage.

Above: Roland-built D.IIa on the Eastern Front with a Pfalz-built D.II in the right background.

Above: Roland D.IIa(Pfal) *Silberfisch* dismantled for transport.

Below: Over-turned D.II with unusual half-circle marking.

Above: Roland D.II in a structural test rig at the factory.

Above: Future ace *Lt.* Hans Pippart flying his Roland D.II(Pfal) while with *Flieger-Abteilung* 220 on the Eastern Front.

Below: *Vzfw.* Richard Scholz in his Roland D.II at *Flieger-Abteilung* 21 on the Eastern Front. The instrument mounted above the altimeter appears to be a clock.

This Page: Scenes of Roland D.II fighters at Brest-Litovsk on the Eastern Front. The German Army captured the city in 1915 and at the time of these photos it was a center for aircraft distribution and aviation training behind the front. At right is a hangar with Roland D.II / D.IIa fighters and Albatros C.III reconnaissance aircraft. Inset clearly shows "W" in two colors on a D.IIa. The air combat environment in the east was much more permissive than the Western Front, which had a much higher density of first-rate combat aircraft. This enabled the Germans to find useful service for mediocre aircraft like the Roland D.II and D.IIa.

Left, Below Left, & Below: Prince Leopold of Bavaria (1848–1930) inspects a Roland D.II or D.IIa during flight operations at Brest Litovsk. By the time of these photos Leopold was the Supreme Commander of the German forces on the Eastern front (*Oberbefehlshaber Ost*). The thick cross insignia indicates the aircraft is Roland-built, and it has the letter "W" painted on the rear fuselage in two colors.

Above: A Roland-built Roland D.II runs up its engine in preparation for take off.

Below: the same Roland-built Roland D.II starts is take-off run.

Roland D.III

Above: The Roland D.III prototype used a conventional cabane structure to further improve the pilot's view forward. This photograph was taken October 26, 1916 at the factory. This first prototype D.III was essentially the same as the earlier D.II except for the center section. Unfortunately, the maneuverability and handling qualities were not improved and few Roland D.III fighters served operationally. Production D.III fighters had a reduced-chord lower wing and larger fin.

The Roland D.II needed further improvement to the pilot's field of view, and the Roland D.III was developed from the D.II to finally solve the problem. The D.III received a robust but conventional-looking cabane structure giving the pilot better visibility forward. The wings were also staggered and the lower wing had reduced span and chord, further improving visibility downward. One D.III flight-tested a number of wings using different airfoils.

Like the D.II/D.IIa, different D.III fighters used both the 180 hp Argus As.III engine and the 160 hp Mercedes D.III, apparently without a different designation to distinguish between them. All D.III fighters were armed with the standard German fighter armament of twin synchronized LMG 08/15 machine guns buried in the forward fuselage. In May 1917 the D.III passed its type test, but the first order had already been given to Roland in February.

Apparently Pfalz built about 30 Roland D.III fighters, 1330–1359/17, under license although it is not known if all were completed. Then the contract was modified for Pfalz to build its own design, the Pfalz D.III, in place of the Roland D.III to finish out the contract using the remaining serial numbers.

The Pfalz D.III fighter design was based on Roland's *Wickelrumpf* "wrapped fuselage" structural technology. The Pfalz D.III, powered by the same 160 hp Mercedes D.III engine, had better maneuverability and flying qualities than the Roland D.III. Furthermore, the poor altitude performance of the Argus engine used in many Roland D.III fighters also compromised their performance and only a limited number of Roland D.III fighters were built. In fact, in June 1917 Roland was given a contract to build Pfalz D.III fighters under license, but the contract was cancelled before any were produced.

Roland D.III Production			
Order	**Qty**	**Serials**	**Notes**
Feb. 1917	50	700–749/17	Roland order; Argus engine
May 1917	50	3000–3049/17	Roland order; Mercedes?
June 1917	100	1330–1429/17	Pfalz Order

The first production Roland D.IIIs did not become available until late September 1917, and by December 98 were on hand. The limited number of Roland D.III fighters that reached the front went to the Eastern Front and Bulgaria, where the air combat environment was much less demanding than the Western Front. In Bulgaria the Roland D.III was viewed as ineffective against the British SE.5a and the Spad 13, but one did manage to down a BE.2.

Roland D.III Specifications		
Engine:	160 hp Mercedes D.III or 180 hp Argus As.III	
Wing:	Span Upper	9.50 m
	Span Lower	8.91 m
	Chord Upper	1.58 m
	Chord Lower	1.20 m
	Gap	1.35 m
	Area	19.83 m^2
General:	Length	6.93 m
	Height	2.9 m
	Empty Weight	717 kg
	Loaded Weight	961 kg
Maximum Speed:		175 kmh
Climb:	1000m	2.5 min
	2000m	6 min
	3000m	10 min
	4000m	15.5 min
	5000m	24 min

Above: The first Roland D.III protoype was powered by a 160 hp Mercedes D.III engine. The wings and tail were the same as the Roland D.II, but production D.III aircraft had a lower wing of reduced chord and a arger fin. Production D.IIIs were powered both by the 160 hp Mercedes D.III and by the 180 hp Argus As.III, apparently with the same type designation.

Above & Below: Front and rear views of the first Roland D.III protoype powered by a 160 hp Mercedes D.III engine.

Right: The structure of the second version of the Roland D.III prototype photographed April 20, 1917. The fin was larger than that of the D.II and first D.III prototype.

Roland D.III first prototype.

Roland D.III 700/17, second prototype.

Roland D.III, unit unknown.

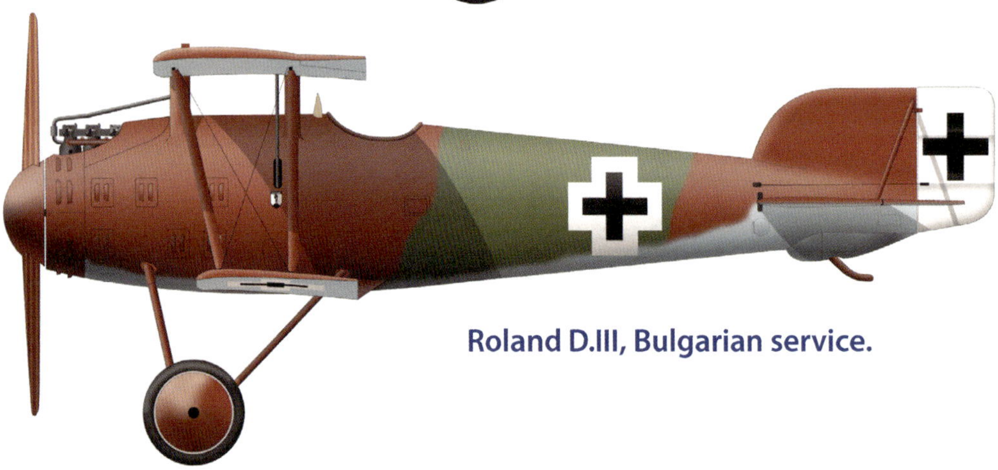

Roland D.III, Bulgarian service.

Above: The second Roland D.III prototype 700/17 (work #2500) photographed on May 16, 1917 shows important changes from the first prototype. The engine is now a 180 hp Argus As.III and the fin is larger for improved directional stability. The chord of the lower wing has been reduced to further improve the pilot's downward field of view, resulting in the lower wing spars being closer together as indicated by the fact the interplane struts now are closer at the bottom instead of being parallel.

Below: The second Roland D.III prototype now carries an anomometer on the interplane strut to measure airspeed.

Above: The second Roland D.III prototype 700/17 (work #2500) photographed on May 16, 1917 at Adlershof.

Below: Roland D.III closeup showing the gun-sight and aileron actuating controls. The machine guns were internal.

Above & Below: Manfred von Richthofen flew the second Roland D.III prototype 700/17 on May 16, 1917 at Adlershof. While his evaluation is unknown, it must not have been negative or it seems unlikely that more Roland D.III fighters would have been ordered. In June 1917 Pfalz received an order for 100 Roland D.III(Pfal) fighters.

Above, Below, & Facing Page: This Roland D.III powered by an Argus As.III engine was photographed on June 13, 1917 at Roland's Kasierdamm facility. This aircraft was used to test different airfoils in an effort to improve flight performance. The D.III was the third configuration of the first Roland fighter generation, all part of a continued search for better visibility for the pilot. The D.I, D.II, and D.III all had a pair of synchronized machine guns mounted inside the fuselage for minimum drag, but these were difficult for the pilots to access in flight to clear jams, which occurred frequently. The fighters also shared the same structural technology and the 160 hp Mercedes D.III engine, although many D.II and D.III fighters used the 180 hp Argus As.III as the Mercedes was in great demand for many other aircraft types. Flight performance, maneuverability, and handling qualities for these very similar fighters were also nearly identical.

99

Above: Roland D.III fighters on Levunovo airfield in 1918; the late-style insignia verify the date.

Left: Bulgarian airmen in front of a Roland D.III of the Bulgarian air service. This aircraft has a 160 hp Mercedes D.III engine. At least six Roland D.III fighters were supplied to Bulgaria. Postwar one D.III, 3000/17, work number 2700, also powered by a Mercedes, was inspected by the British Armistice Commission.

Above: Bulgarian Roland D.III that has been crashed, probably the result of another landing accident.

Above: This Bulgarian Roland D.III has been extensively camouflaged to hide it from Allied aerial reconnaissance.

Above: Roland D.III fighter being assembled outside the LFG flight hangar at Adlershof.

Below: Pilot's view from the cockpit of a Roland D.III illustrates the very robust cabane struts. The small wheel likely adjusted airflow through the offset airfoil radiator, allowing the pilot to control engine temperature.

Above & Below: Roland D.III fighters with their engines removed stored at Leignitz, today Legnica, Poland. Few Roland D.III fighters were built and fewer still saw combat; many, perhaps most, were sent directly to storage.

LFG Roland D.IV / Dr.I

Originally designated the D.IV, Roland's only triplane fighter was re-designated the Dr.I when *Idflieg* promulgated the 'Dr' class designation for *Dreidecker* (triplane). This was the first Roland fighter to use the *Klinkerrumpf* fuselage construction (clinker-built or lapstrake like a boat hull). Power was the 160 hp Mercedes D.III and armament was to be the standard two synchronized guns. The prototype crashed at the end of September 1917 and was rebuilt using a Roland D.VI fuselage, but further testing led to the design being abandoned.

Roland D.IV / Dr.I Specifications

Engine:	160 hp Mercedes D.III	
Wing:	Span	9.45 m
General:	Length	7.32 m
	Empty Weight	930 kg
	Loaded Weight	1208 kg
Maximum Speed:		155 km/h
Service Ceiling:		6,000m

Above & Below: The Roland D.IV / Dr.I used *Klinkerrumpf* fuselage construction to save construction time.

Roland D.V

Three prototypes of the Roland D.V fighter were built in late 1917 with the intention of eliminating the shortcomings of the Roland D.III. Structurally, it was the last Roland fighter to use the *Wickelrumpf* structural technology, and used the complete wing cellule of the D.III. Two synchronized LMG 08/15 machine guns were carried.

One prototype retained the 180 hp Argus As.III, but the other two used the 160 hp Mercedes D.III engine. Limited by the D.III wing cellule, testing revealed little advance over D.III performance and development was abandoned in favor of the more advanced Roland D.VI design.

Roland D.IV / Dr.I Specifications	
Engine:	160 hp Mercedes D.III or 180 hp Argus As.III

Above & Left: The Roland D.V was the last Roland fighter to use the *Wickelrumpf* fuselage construction. It used the D.III wing with lowered front fuselage for improved pilot visibility and refined fuselage for improved performance. The improvement was not enough and the first generation of Roland fighters was finally abandoned.

Roland D.VIa & D.VIb

Above: The Roland D.VI prototype was the 1,000th Roland-built aircraft. Powered by the proven 160 hp Mercedes D.III, it was the protoype for the D.VIa. Another D.VI powered by the 185 hp Benz Bz.IIIa was the prototype for the Roland D.VIb.

The first of three prototypes of the Roland D.VI fighter was rolled out on 17 October 1917, the 1,000th aircraft built by LFG. The D.VI, which used the new *Klinkerrumpf* fuselage construction, was a completely new design. The first D.VI entered flight testing in November 1917 powered by a 160 hp Mercedes D.III. Because the Mercedes was in great demand, especially for fighters, at least two D.VI prototypes were powered by the 185 hp Benz Bz.IIIa. Two synchronized LMG 08/15 machine guns were carried.

The D.VI participated in the First Fighter Competition in January 1918, and the D.VI passed its type test on 9 February 1918. Based on its promising performance in the fighter competition, a small production batch of 50 D.VI fighters was ordered. Subsequently, *Idflieg* placed orders for a total of 350 D.VIa and D.VIb fighters. Production was 150 D.VIa fighters powered by the 170 hp Mercedes D.IIIa engine, and 200 D.VIb fighters powered by the 185 hp Benz Bz.IIIa engine. Both variants reached combat units in May–June 1918.

The Roland D.VIa was slightly faster and more maneuverable than the Albatros D.Va and Pfalz

Roland D.VIa Specifications

Engine:	170 hp Mercedes D.IIIa	
Wing:	Span Upper	9.40 m
	Span Lower	8.68 m
	Chord Upper	1.40 m
	Chord Lower	1.20 m
	Gap	1.49 m
	Area	21.7 m^2
General:	Length	6.40 m
	Height	2.60 m
	Empty Weight	655 kg
	Loaded Weight	845 kg
Maximum Speed:		190 km/h
Climb:	1000m	2.5 min
	2000m	6.0 min
	3000m	11.0 min
	4000m	18.0 min
	5000m	25.0 min

Roland D.VIb Specifications

Engine:	185 hp Benz Bz.IIIa	
Wing:	Span Upper	9.40 m
	Span Lower	8.68 m
	Chord Upper	1.40 m
	Chord Lower	1.20 m
	Gap	1.49 m
	Area	21.7 m^2
General:	Length	6.40 m
	Height	2.60 m
	Empty Weight	655 kg
	Loaded Weight	845 kg
Maximum Speed:		200 km/h
Climb:	1000m	2.5 min
	2000m	4.9 min
	3000m	7.8 min
	4000m	12.3 min
	5000m	19.0 min

Above: Another view of the first Roland D.VI prototype; it was completed the morning of 27 Oct. 1917.

Below: The second Roland D.VI prototype featured a redesigned fin and rudder in a search for improved handling and maneuverability.

Roland D.VI Production Orders			
Ordered	**Qty**	**Serials**	**Notes**
Feb. 1918	3	220–222/18	Prototypes
Feb. 1918	50	1200–1249/18	D.VIa
Apr. 1918	50	2212–2261/18	D.VIb
June 1918	50	3600–3649/18	D.VIa
June 1918	50	5000–5049/18	D.VIa
Aug. 1918	50	5600–5649/18	D.VIb
Sep. 1918	50	6100–6149/18	D.VIb
Sep. 1918	50	7500–7549/18	D.VIb

Above & Below: The second Roland D.VI prototype featured a redesigned fin and rudder; however, like the first prototype it was powered by a 160 hp Mercedes D.III engine.

D.IIIa that it was intended to replace, and was structurally stronger than the Albatros V-strutters with single-spar lower wings. However, its climb rate was less and the *Jasta* pilots viewed it as a marginal improvement over the earlier types. The more powerful D.VIb, the only production aircraft powered by the 185 hp Benz Bz.IIIa, was faster and had better climb than the earlier Albatros and Pfalz fighters. A good fighter, the D.VIb was, like all other types, over-shadowed by the Fokker D.VII, and it was the last production Roland fighter. However, limited D.VIa/D.VIb production lasted until the war's end.

This Page: The second Roland D.VI prototype featured a redesigned fin and rudder with aerodynamic balance. The first three D.VI prototypes each had a different fin and rudder design in a search for better flying qualities.

Above & Below: The third D.VI prototype flew at the First Fighter Competition. This prototype has an under-fuselage fin. Powered by the 185 hp Benz Bz.IIIa engine, it was the prototype of the Roland D.VIb. In addition to the different tail designs, the different D.VI prototypes tried both aerodynamically balanced and unbalanced ailerons and elevators, all in pursuit of the best maneuverability and handling qualities. All production D.VI aircraft settled on aerodynamically-balanced ailerons and rudders. The Mercedes-powered D.VIa and some early-production D.VIb fighters had unbalanced elevators, but aerodynamically-balanced elevators were standardized for the D.VIb.

Above & Below: D.VIb 2217/18 photographed on June 10, 1918, was used for a series of tests to investigate an extremely flat wing profile. Interestingly, the new wings were supported by four I-struts.

Above: Another photograph of D.VIb 2217/18 taken on June 10, 1918, during its use for a series of airfoil tests.

Above: A dark-painted D.VIa of *Jasta* 32. The aircraft was painted at the unit.

This Page: A Roland D.VIa of *Jasta* 23b photographed postwar at Koblenz, where it was evaluated by Allied pilots. The white circle on the black fuselage band is thought to have been added after it was given to the US Air Service as part of the Armistice requirements.

Left: Roland D.VIb 6141/18 crashed by German pilot *Lt.* Hans Christian Donhauser of *Jasta* 17 on January 1, 1919 at Coblenz. He walked away, but was later killed in a crash during a test flight. Such accidents were regarded as normal at the time.

Below: A Roland D.VI from *Jasta* 23b postwar with an American pilot in the foreground.

Above: Roland D.VI at *Armee Flug Park* 17.

Below: An unidentified D.VIa with a painted fuselage. A Fokker D.VII rests in the background.

Above: A group of Americans poses with late-production Roland D.VIb 7504/18 at Koblenz in 1919.

Below: A spectacularly-painted Roland D.VIa of *Jasta* 32b; *Uffz*. Krautheim at left with *Uffz*. Marchner next; the other men are unknown.

Facing Page: *Lt.d.R.* Emil Koch of *Jasta* 32b with his colorful Roland D.VIa. Koch commanded *Jasta* 32b from July 23 to October 24, 1918, when he was wounded while strafing advancing British troops.

This Page: *Lt.d.R.* Emil Koch of *Jasta* 32b ready to fly his Roland D.VIa. Koch was an ace who scored 7 victories.

Above: Roland D.VIa 3612/18 was photographed at Koblenz postwar after it was delivered to the Americans. The aircraft was previously flown by *Lt.* Hans Jungwirth of *Jasta* 78b.

Right: Roland D.VIa with its pilot in the cockpit ready for a mission. The four vents on the starboard nose were an indication of the Mercedes engine being used. See also the photos of *Lt.* Koch's D.VIa on preceding pages.

Left: The first Roland D.VIa fighters arrived at *Jasta* 35b on May 14, 1918. This early-production D.VIa (identified by its straight metal interplane struts faired with wood) was damaged by Albatros D.Va 7186/17 (in the background) while being flown by *Lt.* Bomer on June 1, 1918. The chevron on the top of the upper wing was a unit marking.

Roland D.VI

Roland D.VIa of ***Lt***.d.R. Emil Koch, *Staffelführer* of *Jasta* 32b. Koch was an ace who scored 7 victories.

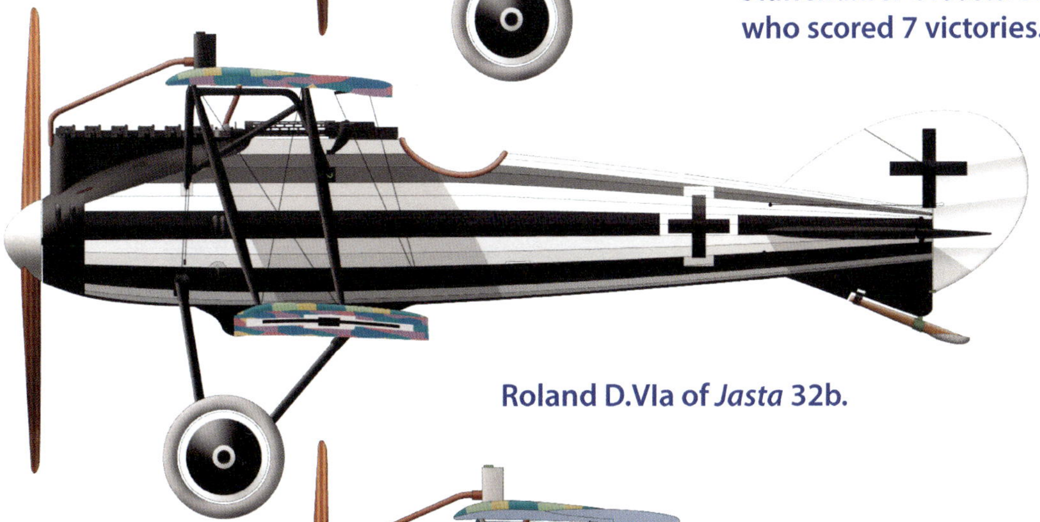

Roland D.VIa of *Jasta* 32b.

Roland D.VIa 3612/18, ***Lt***. Hans Jungwirth, *Jasta* 78b.

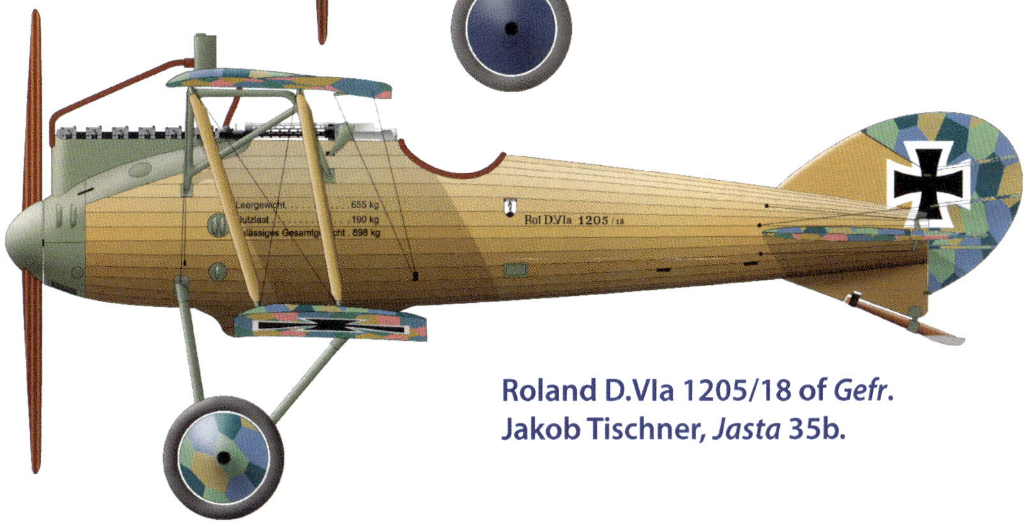

Roland D.VIa 1205/18 of *Gefr*. Jakob Tischner, *Jasta* 35b.

Roland D.VIa of ***Lt***. Otto Kissenberth, *Jasta* 23b. Kissenberth scored 20 victories and was awarded the *Pour le Mérite*.

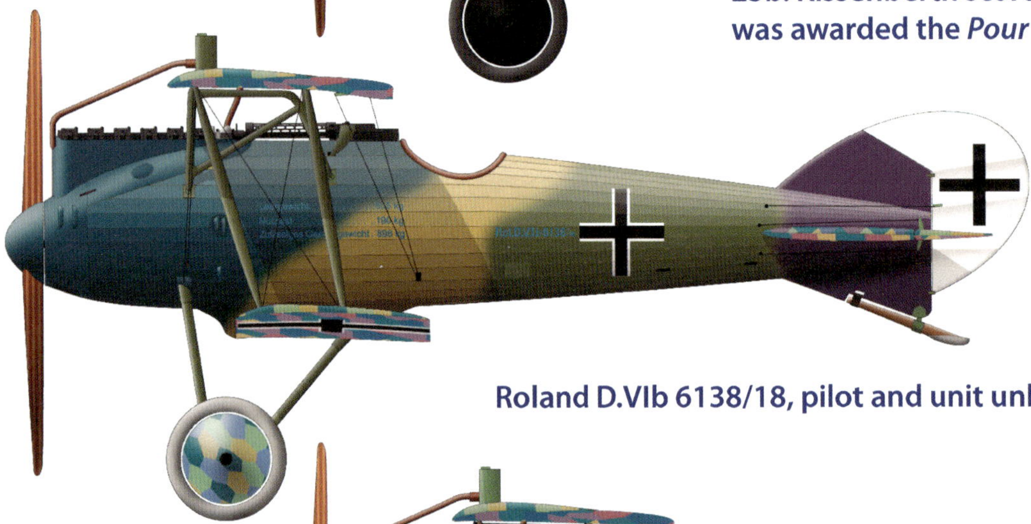

Roland D.VIb 6138/18, pilot and unit unknown.

Roland D.VIa of *Jasta* 32b.

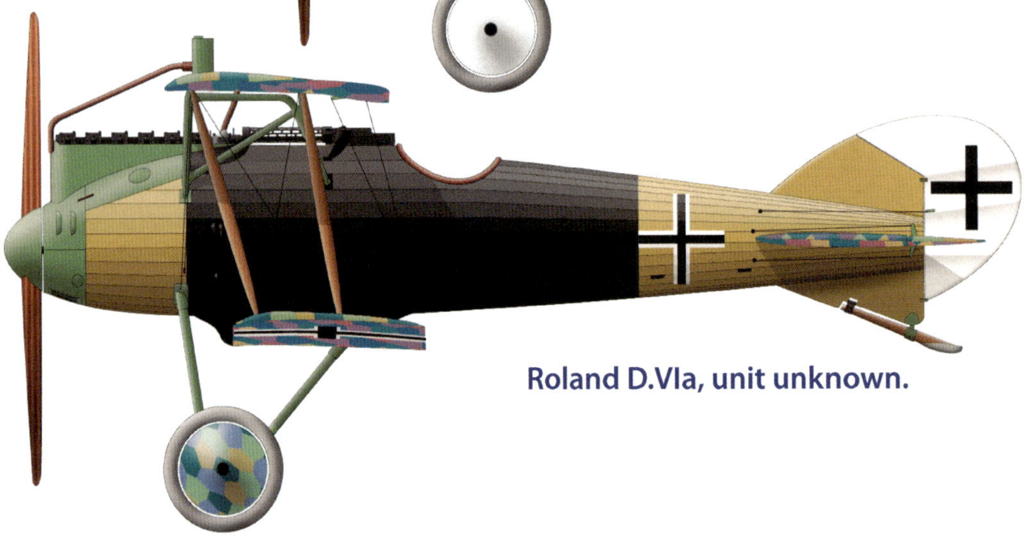

Roland D.VIa, unit unknown.

Above: The spectacular *Edelweiss* marking on *Pour le Mérite* ace Otto Kissenberth's Roland D.VIa of *Jasta* 23b.
Below: Roland D.VIa 1219/18 was the 20th production aircraft and reportedly served with *Jasta* 30.

Above: Aircrew of *Jasta* 35b gather for an informal portrait with a Roland D.VI. Rudolf Stark, sitting on the tail in his swim trunks, was the *Jastaführer*. The chevron on the top wing was part of the unit markings.

Below: Roland D.VIb 6117/18, with a painted nose and spinner, was an aircraft of the sixth production batch. Here it is in Allied hands after the Armistice.

Above: Roland D.VIa 1211/18 was the 12th aircraft in the first production batch. Despite being from the first 20 aircraft built, which are all thought to have the straight metal interplane struts, this aircraft now appears to have the all-wood interplane struts normally fitted to later-production aircraft.

Above: Roland D.VIa 5012/18, an aircraft from the 4th production batch, was sent to Japan as war reparations and was photographed upon arrival.

Above: This closeup of a disassembled Roland D.VIb shows details of the cockpit, guns, and engine.

Above: Roland D.VIa 5023/18, an aircraft from the 4th production batch, with Lottchen in the cockpit.

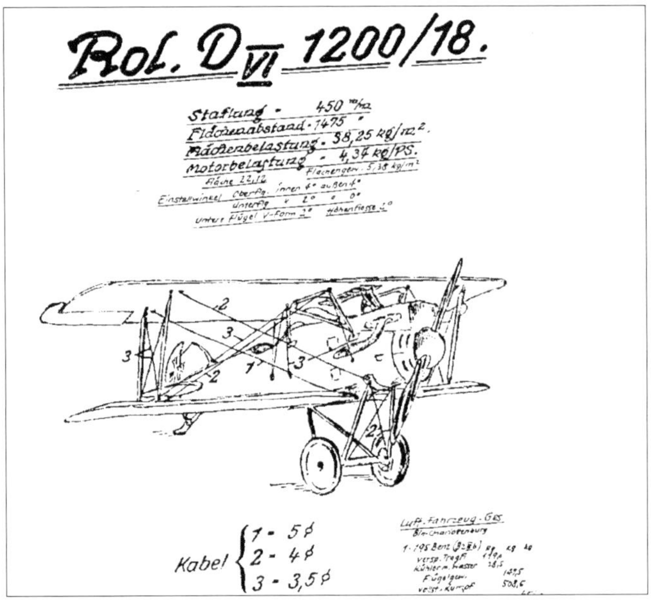

Right: Roland D.VI rigging diagram.

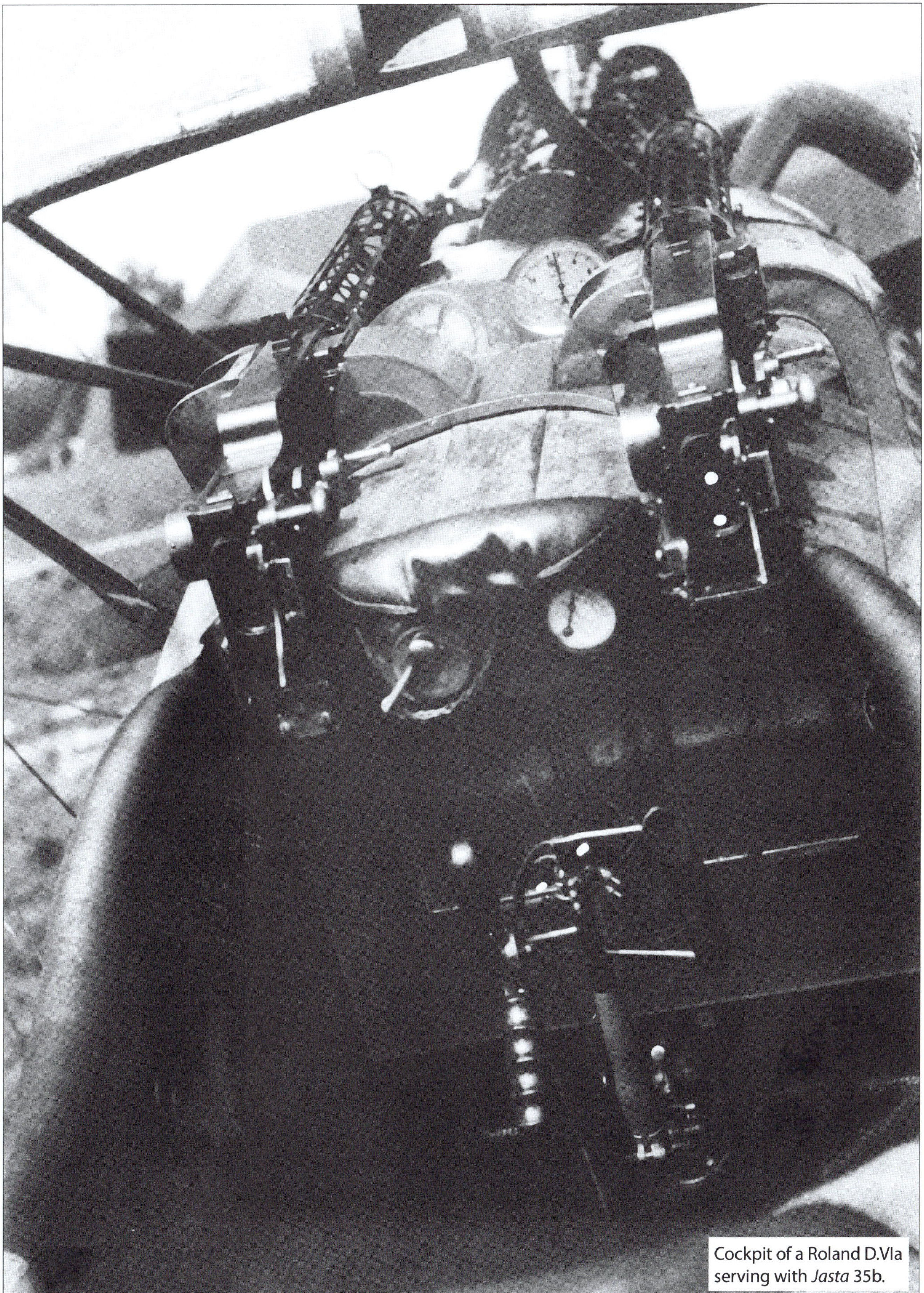

Cockpit of a Roland D.VIa serving with *Jasta* 35b.

Right & Bottom: Roland D.VIa 12xx/18. The fuselage insignia has been over-painted. The men and unit are unidentified.

Right: Closeup of the data painted on the fuselage of Roland D.VIa 12xx/18.

Above: At left, a Roland D.VIa of *Jasta* 35b with an Albatros D.V or D.Va on the right.

Left & Below: Roland D.VIa 1223/18 served with *Jasta* 35b. Mostly in factory finish and markings, the chevron on top of the upper wing was part of the unit markings. *Jasta* 35b received its first five Roland D.VIa fighters from *Armee-Flugpark* 17 on May 7, 1918 and did not start receiving Fokker D.VII fighters until August 24, 1918. The Roland D.VIa was more maneuverable than the Albatros fighters the *Jasta* had used until then but otherwise offered little performance improvement. However, it was stronger than the Albatros and not subject to wing failure.

Above: Roland D.VIb 2216/18 undergoing evaluation at Adlershof in April–May 1918. This aircraft is from the second production batch, which was the first batch of D.VIb aircraft. According to official *Idflieg* drawings the D.VIb was supposed to receive an aerodynamically-balanced elevator, but this photo proves that was not fitted to all production D.VIb aircraft.

Below: This appears to be another photograph of Roland D.VIb 2216/18 shown above. The Roland D.VI had clean lines. The lower wing was attached to a pylon below the fuselage to provide sufficient wing gap to prevent airflow interference between the wings.

This Page: Roland D.VI in Czechoslovakian service postwar. Czechoslovakia was apparently the only other country other than the USA to use the Roland D.VI postwar. A small but unknown number were used.

Above: Roland D.VI with smooth, apparently monoque fuselage; no details of this interesting modification are available.

Left: Roland D.VIb 2253/18 was flown in 1919–1920 by Czechoslovakia.

Right: Crash of Czechoslovakian Roland D.VI number "2".

Above: Americans starting Roland D.VIb 6142/18 at Romorantin postwar.

Below: Roland D.VIb 6138/18 was an aircraft of the sixth production batch. Here it is stored by Americans after the Armistice. These late-production aircraft were finished like the Halberstadt CL.IV(Rol) in the NASM.

Above: Roland D.VIb 6142/18 disassembled and stored at Romorantin postwar.

Above: Roland D.VIb 6142/18 ready for flight at Romorantin postwar.

Above: Roland D.VIb 6142/18 taxiing for takeoff at Romorantin postwar. This late-production aircraft was finished in the same colors that Roland used on their license-built Halberstadt CL.IV(Rol) aircraft. This is significant because one of these aircraft has been carefully restored and is now on display at the Udvar-Hazy facility of the National Air & Space Museum as shown at right.

Right: Restored Halberstadt CL.IV(Rol) at NASM's Udvar-Hazy facility. The factory finish of this Roland-built aircraft was well documented during its restoration. Roland finished late-production Roland D.VIb aircraft from the 6100–6149/18 and 7500–7549/18 production batches in the same colors. Some of these aircraft are shown above and on the facing page.

This Page: A Roland D.VIb, likely 6142/18, at Romorantin postwar.

Above & Right: Roland D.VIb 6142/18 at Romorantin postwar. This aircraft was photographed a number of times.

Left: Roland D.VIb, probably 6142/18, being readied for a test flight at Romorantin postwar.

P
132
7502

P
132
7502

This Page & Facing Page: Roland D.VIb 7502/18 from the final Roland D.VI production batch was brought to America after the war. Later it was used for a series of camouflage experiments by the U.S. Army as shown here.

Right: Roland D.VIb 7504/18, from the same production batch as its sister 7502/18, is shown postwar at Koblenz. As originally built by Roland it has the same factory finish as other late-production D.VIb aircraft shared with the Roland-built Halberstadt CL.IV(Rol) restored by NASM.

Above & Facing Page, Top: The US Army Air Service used the Roland D.VIb as one of a number of aircraft in day camouflage experiments in 1921. For the full story of this interesting facet of the D.VI's career see Colin Ower's article in *Windsock*, Vol.29 No.2.

Roland D.IV/Dr.I prototype triplane fighter, initial version before crash and rebuilding.

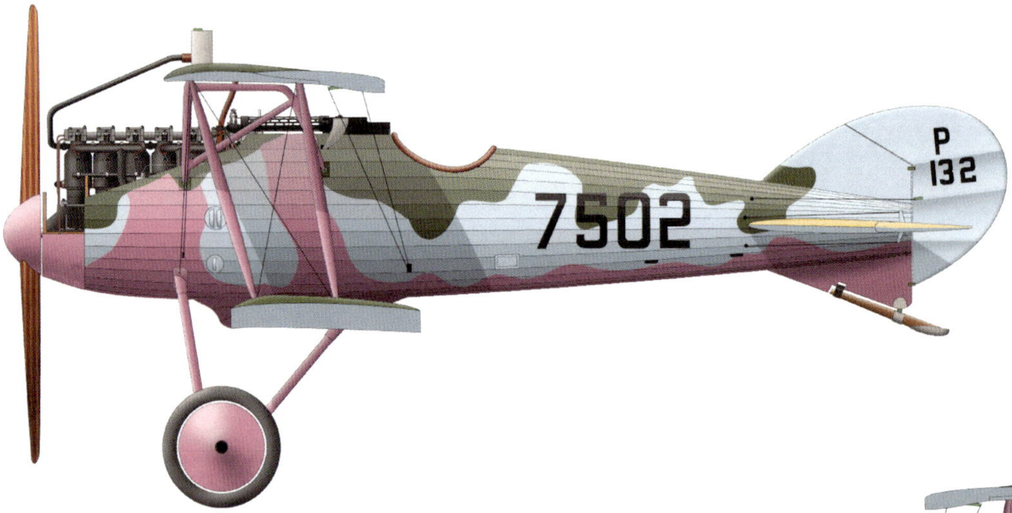

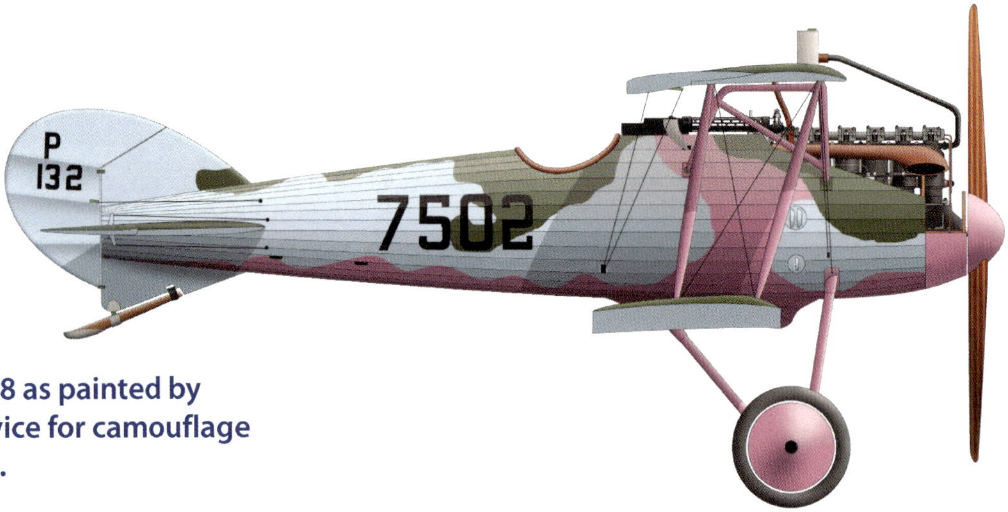

Roland D.VIb 7502/18 as painted by the US Army Air Service for camouflage experiments in 1921.

This Page: The fuselage of Roland D.VIb 2225/18 in Poland.

This Page & Facing Page: The only remaining aircraft of Roland design in the world is this D.VI fuselage in Poland.

Above: Disassembled and packed Roland D.VI fighters on the train on their way to the front.

Below: Late-production Roland D.VI aircraft stored at Romorantin postwar. These aircraft were to be shipped to America, and some were. With all these pristine aircraft available, how is it that none were preserved in a museum?

Above: Late-production Roland D.VI fighters in storage at Romorantin postwar.

Right: A Roland D.VI in Ottawa; America was not the only recipient of war-booty Rolands postwar, nor the only country to fail to preserve one for future generations.

Below: An American pilot taxis a Roland D.VIb, perhaps 7504/18, for takeoff at Romorantin postwar.

Roland D.VII & D.VIII

Above: The Roland D.VII prototype 224/18 was powered by the experimental Benz Bz.IIIbo ungeared V-8.

The Roland D.VII was developed in parallel with the D.VI; the key difference was the engine. The D.VII, which also used the *Klinkerrumpf* fuselage construction like the D.VI, was powered by an experimental 185 hp Benz Bz.IIIbo direct-drive V-8 engine. The D.VII had two synchronized LMG 08/15 machine guns and also participated in the First Fighter Competition.

During flight testing the ailerons were revised and D.VI-type vertical tail surfaces were introduced. The D.VII had some good qualities, including excellent field of view for the pilot. The D.VII also participated in the Second Fighter Competition in May 1918, but the experimental engine was troublesome and not ready for production and D.VII development was finally abandoned.

Roland D.VII Specifications

Engine:	185 hp Benz Bz.IIIbo V-8	
Wing:	Span	8.84 m
	Area	20.83 m^2
General:	Length	6.10 m
	Height	2.80 m
	Empty Weight	666 kg
	Loaded Weight	858 kg
Maximum Speed:		190 km/h
Climb:	5000m	16.2 min

Right: Different tail surfaces were tried on the Roland D.VII prototype but the aircraft was doomed by its troublesome engine. Development of the similar Roland D.VIII, powered by the geared version of this engine, was likewise terminated due to engine problems.

Above & Below: The Roland D.VII was very similar to the prototype D.VI except for using the experimental 185 hp Benz Bz.IIIbo direct-drive V-8 engine. The experimental engine was unreliable and not ready for production, dooming the D.VII.

30323449R00116

Made in the USA
Lexington, KY
27 February 2014